The Stages of a Grounded, Healthy Gestalt Figure

Frontispiece. A generalized chart depicting stages of unimpeded, free-flowing figure development, resolution and involution.

Phase of engagement and action of the figure in the environment.

Energy scale (ψ/log)

CONSUMMATING ACTION
To meet needs and solve problems
Point of ENGAGEMENT with the self-nurturing, need satisfying action particularly with other people.
Apotheosis, Apogee.

<5>

<4>   <6>

ENGAGEMENT ENDS AS NURTURING ACTION COMPLETES AND NEEDS, ETC., FROM STAGE <1> ARE MET.
Recognition that the particular desire, appetite or interest is now satisfied and the tension resolved.
Whatever was lacking is now provided.
Good bye to companions.

MOBILIZATION OF RESOURCES.
CHOICE AMONG OPTIONS
Contact with the need, etc., satisfying process, particularly communication with a partner.

<3>

EXCITEMENT, AFFECT, PRELIMINARY ENERGY,
Intention known about satisfaction of needs.
Thoughts about options.
Rising interest enthusiasm and animation.

<7>

AWARENESS OF AND ENJOYMENT OF ENDING, FULFILMENT, SATIATION, SATISFACTION, PLEASURE AND GRATIFICATION
Endeavour is rewarded with warm feelings, particularly concerning the companion now gone.
Gives thanks to self for a job well done.

Phase of growing figure and consciousness

All stages are choice points; go on be stuck or back off.

ALERT AWARENESS AND VERBALIZED THOUGHTS ABOUT THE PROBLEM <2>
Expression of the need for nurture, want, intention, interest, concern or desire.

INITIAL AFFECT AND/OR NEED, WANT, APPETITE, IMPULSE, CURIOSITY OR ARISING HUNGER. <1>
A felt-protofigure?
Non-verbal recognition of tension, of something lacking and/or required.

WITHDRAWAL
The end of this figure.
"Pleasant afterglow."
Closure to an equilibrium state without pressing needs, etc., and probably to concern with a new self-nurturing figure.

<8>

External stimuli

<0>
protogestalt matching

Internal Stimuli plus memories of previous similar figures

NEUROLOGY
= mind/brain processes

<0'>

These new events are remembered and form new protogestalts

time

SCIENTIFIC GESTALT

AN INTRODUCTION TO PROTOFIGURE THERAPY

RAY EDWARDS PH.D., M.A., B.SC.,
FELLOW OF THE ROYAL SOCIETY OF MEDICINE.

There came Red Stallion from the East
Who said:
Ride me with courage and destiny

Rainbow from the West
Said:
Yes; The time has come.

From the North the hollow, reverberant voice
Said:
Preserve me from Injury

Sun sustenance came from the South.

authorHOUSE®

AuthorHouse™ UK Ltd.
500 Avebury Boulevard
Central Milton Keynes, MK9 2BE
www.authorhouse.co.uk
Phone: 08001974150

Published by AuthorHouse 8/13/2012

ISBN: 978-1-4772-1418-3 (sc)
ISBN: 978-1-4772-1419-0 (e)

Treetops
Old Lyme Hill
Charmouth
Dorset
DT6 6BP

☎ 01297 560 291
Skype = raywhedwards

raywhedwards@gmail.com

Scientific Gestalt: An Introduction to Protofigure Therapy

Contents

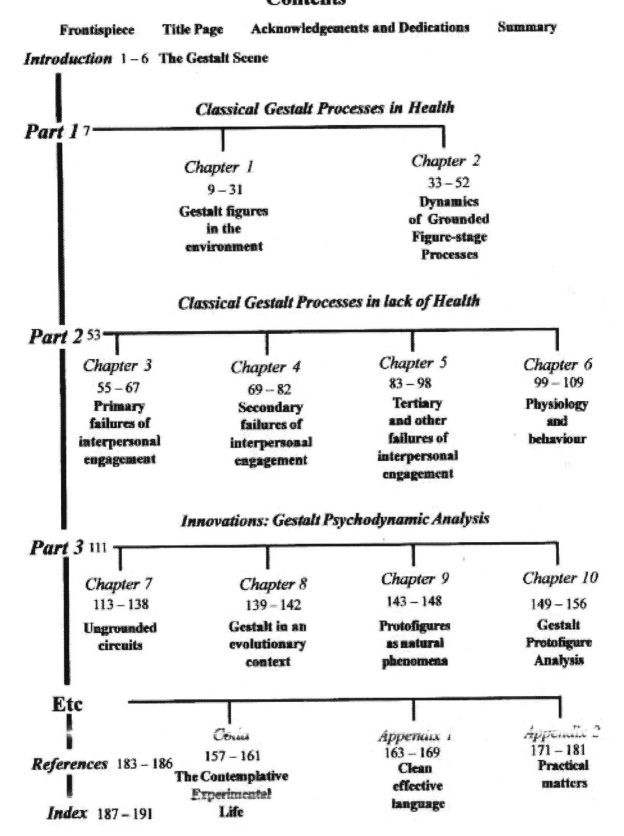

Acknowledgements and dedications

To my many teachers, those who spoke for me to hear and those who wrote for me to read. Particularly to Ischa Blomberg who knew what elephant shit was, to Ursula Fausset who knew what love was, to Petrûska Clarkson who knew what a cycle was, to Margaret Turpin who taught me how to transact, to Maria Gilbert who knew what laughter was, to Marriane Fry who knew and to Malcolm Parlett who edited.

To Marvin Kaplan of the Gestalt Institute of Cleveland from whom I learned the "feel" of being a gestalt therapist.

To Maxwell Cade who was happy when I went into "state five" with no bother and taught me to create a haiku to match each occasion.

To Penny Edwards, (Née Vinter, 1927 → 1989) a very different person with very different ideas and experiences; we lived together and we worked so well together when running couples groups.

To Zara and Christopher for not objecting to living with Penny and me and the next generation, Kamille, Reece and Alexander for much fun.

To Andrew, John, Alex, Arthur and many others for shared love.

To John Southgate whose views on dialectical counselling reached me via Penny,

To Dave Dobson who wondered a lot and David Grove who wondered what happened next and discovered protofigures without knowing it.

To the many citizens of Camden Town, Hampstead and Highgate who sat with me in group exploring the dynamics of interpersonal relationships. And to others who made a pair with me there and in Dorset

To the many adult students who, over many years, listened to me and read what I wrote. These students were in the Westminster and Camden Adult Education Institutes, in the City University and in the Mental Health Nurse Training School at the Romford College. We contacted and got into the action of sowing seeds for one another. Our fields were indeed fertile.

To the nerds who invented www.* and *.com

To Jean Sibelius for *En Saga*.

Scientific Gestalt:
An Introduction to Protofigure Therapy

By Ray Edwards

Summary

This book concerned with *Scientific Gestalt* introduces *Protofigures* and their use in therapy, first placing the subject matter in its historical context in the **Introduction**. The early workers, concerned with gestalt phenomena, were academic psychologists who showed little interested in the human predicament. Emphasis in this book is on people and much abbreviated "case studies" will be presented to illustrate arguments. **Part 1** is concerned with observations obtained with healthy people and in intact situations: classical, Perlsian gestalt.

Chapter 1 deals with gestalt ideas, principally recognizing perception of each figure in relationship to a ground and the consequences of these observations in relationship to theories of perception, needs, action and boundaries with the environment (field). A hypothesis is presented concerning the precursor of each gestalt figure showing that it exists in the neurological matrix as a module, a protofigure.

Chapter 2 is concerned with the observations made in the middle of the 20th century that showed that subdivisions of figures can be detected as the gestalt grows and develops and these take the form of discrete stages of perception; awareness, excitement, affect, energy expression, consummating activity and withdrawal. These ideas are based on the contributions of F. Perls, Hall and Zinker. Knowledge about these stages is conditional on figure development being slow enough to allow discernment

The stages of figure development are regarded as primary to all other activities and are discussed mainly in terms of one-off sequences. Multiple sequences can appear as recycling processes as with figures exemplified by eating where only minor differences may be experienced between the successive unit circuits.

Part 2 details aspects of gestalt analysis as happens when the person observed is not healthy and/or his situation is not intact, is fragmented, is blocked. It is envisaged that interruptions of gestalt figure development may be seen at three levels. Thus **Chapter 3** deals with what I call primary interruptions of process, failure of figure stage development so that the gestalt is not completed. An example occurs when a timid person fails to muster enough energy to carry out his project, backs off, and is left grossly dissatisfied with self.

In **Chapter 4** the familiar interruption and failure processes described by Fritz Perls are designated as secondary. Here, as an example, a sad person projects his problems on other people who he can then blame for his predicament.

Also discussed are the ideas of Anna Freud concerning the predicaments of self healing which frequently lead to worsened situations.

In **Chapter 5** tertiary aspects are dealt with including, guilt, shame, supurating resentments, vengefulness, blaming and Bernian games.

The physiological state of the body has considerable bearing on behaviour and is discussed in **Chapter 6**.

Part 3 is concerned with innovations in gestalt theory and practice that have occurred since the Perlsian era.

Grounded gestalt figures and their stages of development were described in Chapter 2 and **Chapter 7** introduces the ungrounded forms that occur in depression.

In the present century the subject matter of psychology is being transformed by authors who are interested in the ideas and practices of Darwinian evolution so **Chapter 8** sets the gestalt approach in that context.

Protofigures, in the form of "butterflies in the stomach," "lumps in the throat," "clouds are hanging over me," etc., were introduced in the Introduction and they return again in **Chapter 9** where their role in "seeding" new, curative gestalt figures are presented as "case studies." Careful linguistic control is described as of value in promoting the maturation of the new gestalt figures, stage by stage, and thus in providing relief from the occurrence of the cataclysmic affect that otherwise accompanies gross trauma manifestations, the hyperactivity states, including panic attacks and phobias as well as lesser deleterious events.

The more theoretical aspects of protofigure analysis are presented in **Chapter 10** where I share with the reader some of the personal problems I encountered when setting out to ensure that this would be a fully scientific study.

A **Coda** counterbalances the introduction, summarizes and rounds off aspects of discussion

In **Appendix 1** the special sub-set of language described by Grove and called "clean language," similar to the linguistic approach of Levitski and Perls, is described.

Appendix 2 presents practical procedures that were referred to in the main text but were too extensive for comfortable inclusion there.

—=(☆)=—

Introduction

The Gestalt Scene

The integrated personality has *style*, a unified way of expression and communication ... the person who has *style* does not come for therapy ... The people who want and need therapy are the ones who are stuck with their anxiety, their dissatisfaction, their inadequacies in work and relationships, their unhappiness. (L. Perls[1]) [her emphases].

Certain manifestations indicate that alien-self is in control. It is useful for the therapist to explore these and use them as a guide for therapeutic interventions. (F. Perls[2])

My purpose in setting out this introduction is to stimulate your interest in matters that come later and thus I present many incomplete statements and even repeat some of them.

Scientists of the early 20th century (Goldstein, Koffka, Köhler, Lewin and Wertheimer) recognized a tendency of information to be recognized in whole units and called them gestalts. These people were initially interested in perception and also recognized that each gestalt consisted of an active component, termed the figure, and a context, the ground.

The present study depends on recognition that the gestalt figure is rooted in the ground and that the roots have properties that are important enough to name them in the gestalt context, protofigures.

An introduction to protofigure therapy requires introduction to the new phenomenon of protofigures, how to encounter them in the therapy situation and how to manipulate them for the benefit of people with behavioral and psychological disabilities.

An understanding of protogestalt therapy depends on a scientific examination of general gestalt theory which will

> *Pathology*: Just as a lesion of the skin is a hole through which blood leaks so psychological lesions are holes through which adverse behaviours leak.
>
> The healing task in both circumstances is to regain the whole state, to complete the gestalt.

thus be the subject matter of the first few chapters of this exposition. The protofigure aspect of the protogestalt changes with time, develops and, in health, matures to form a complete curative gestalt.

After a consideration of the gestalt theories of perception, needs and action in Chapter 1 there follows a dissection of the stages of gestalt figure evolution, finding needs, excitement, energy, mobilization of resources, and action to satisfy the needs as consummation of the gestalt.

Gestalts do not always develop to completion because interruptions occur and the pathological effect of these are considered in three chapters concerned with the three logical levels, interruption at the figure stage level, occurrence of projection, etc., and concern with personality problems like shame and blame.

Failure of physiological phenomena have psychological consequences and failure of breathing, eating, excretion and sexual function are considered in Chapter 6, the last chapter concerned with history.

Part 3 presents the innovations of this study. The occurrence of sequences of staged events, rather like the figure stages, that do not ground in depression is described in Chapter 7. Attention to these stages leads to discovery of one stage that is more amenable to therapy than the others, called "the weakest link", and a return to fully functioning, grounded, gestalt formation then occurs.

Chapter 8 sets the current study in a wider context, including neurological factors and a consideration of the effect of Darwinian evolutionary factors, such as the contrasting effects of inherited, genetic and social factors.

Of principle concern, in Chapter 9, are the eponymous protofigures. Here the origin in Grove's study of traumatized patients is acknowledged as the origin of this present study. As indicated above, felt-sense phenomena in the form of "feeling" butterflies in the stomach or lumps there or in the throat, clouds hanging over, walls obstructing passage, etc.

Explanations and examples illustrate how the felt-protofigure develops using Grove's questioning technique and his "clean language". It is hypothesised that this produces a new curative gestalt. A brief discussion of the neurology of the process is provided in Chapter 10.

Credentials

As a scientist I approached the gestalt system with a non-scientific interest. I was intrigued by the artistic ethos – until I encountered patients. The clinical medical attitude I had picked up in medical school provided a down to earth healing environment that contrasted with the fantasy attitude of F. Perls, Simkin, and colleagues and their contemporaries. It took me twenty years to sort out my dilemma, to escape the logic / illogic impasse I had entered by blindly accepting what my gestalt trainers provided for me.

My tutor during the breaking of my impasse was gestalt man David Grove though he could not have known what was happening to me as he cut the affect from the trauma of real patients. This is not a long story though it has much detail – which will be delineated during the rest of the pages of this book.

I have no doubts about that after working in medical science for about 50 years. I have about 50 papers published in learned clinical journals and books. I was elected to fellowship of the Royal Society of Medicine on the basis of the scientific content of these publications – it is unusual for non-clinically qualified persons to receive this accolade. I have lived in the company of medical scientists where I have enjoyed euphoria and passions.

The two cultures

Canutists never stop believing
that personal power can stop the tide running.

Snow's observation[11] concerning the antithesis of the arts and the scientific attitudes distils, in the gestalt context, to form the existential versus the scientific approaches. Most commentators in the gestalt field concentrate on the former aspect so little attention will be given to it here other than to commend Philippson's essay[9] entitled *The Emergent Self; An Existential-Gestalt Approach*.

Characteristics of the Gestalt model

Various writers have used the word *gestalt* in several senses. Originally, according to Cassell's German dictionary (1909) the word meant: form, shape, figure, to form, to fashion, to take shape, to appear, to turn out; and for *gestaltung*, formation, forming, modeling, form, figure, shape, condition, state, phase. It would have been with these meanings that Goldstein, Koffka, Köhler, Lewin, Wertheimer and colleagues thought and wrote about gestalts.

These days, in english, *gestalt* is the name for the study of whole objects or events and it is an aspect of the gestalt, the figure, that has shape, form, and a state of existence. At sometime within the passing century it was recognized that the gestalt figure had an origin in a ground and that it was, when healthy, well grounded. If the gestalt could be seen as an object then that object was set in a ground. *Example*; a mandala (e.g., Chart 3·1) could be a complete entity set in a ground of Buddhist reverence.

Alternatively, if the gestalt was an evolving entity it could be considered to have arisen from a ground and that it would revert to ground when its mission was done. *Example*; A potter creates a vase by first preparing his wheel, oven, clay, etc., and finally tidying away his wheel, oven and residues of clay while proudly showing off his pot.

These two examples exemplify static and dynamic gestalts. In the study of human character and behaviour dynamism is of singular importance and will be the core of study in this book.

Generalizing, it can be seen that all phenomena arise, are born, are created as gestalts and pursue a natural history as such. Lack of health occurs if the processes of the gestalt, its figure development, are interrupted, because it is with the evolution of the figure that the work of the gestalt, the contribution to the life of the person concerned, is done.

Summary of definitions. In the human context a gestalt is a whole entity that occurs to meet a need, to satisfy a curiosity, an appetite or hunger or complete an event. The active part of a healthy gestalt is a figure which develops in time in discernable stages and reaches completion. The gestalt originates in what is traditionally called a ground but which is actually a matrix or module of CNS neuron tissue. This aspect of a gestalt which may be recognized on certain occasions is termed a felt-protofigure or more simply, protofigure.

There are two boundary phenomena for each gestalt, that relating to the CNS and that, as referred to above, relating to the environment. The latter varies with time and demonstrates functional behaviours traditionally recognized as projection, introjection, etc., now thought of as venturing, assertion, aggression, passivity, etc., of potential deleterious effect.

CNS and other nervous system matters are subject of extensive investigation in chapters 2, 9 and 10.

The characteristics of a scientific approach to gestalt phenomena

A good theory is an
effective guide to action.

Lewin

Scientists have always been practical people, running experiments to discover natural facts and then generalizing to form theories or having a theory and devising experiments to test validity. If the experimental results were nearly valid the theory was modified to make it fit, followed by further validation experiments. This iterative approach might continue for some time until the observer(s) was/were satisfied.

Although initially one person may have made an observation, repetition by other people produced the same result, wherever the experiment was done. Such observations could be in the style of measuring the percentage of oxygen in our air – it is always 21% ± a small error.

I have also the propositions of Kuhn[12] and Popper[13] to consider and have no quarrel with them. If the present observations are recognized as valid by my peers they will at least appear in the histories of psychology.

That is enough about science for now – more will appear after an introduction to the gestalt approach and a discussion of the relationship of the experimenter and the facilitator.

Ethical relationships

Anna Freud[4] emphasized
that people want happiness,
and want to avoid anxiety
and all negative emotions.

The relationship of the experimenter and the facilitator is maintained mostly by dialogue. The experimenter used to be referred to as the patient in the medical model or the client in the private practice model. To celebrate the uniqueness of this person he or she is referred to here as the experimenter because the objective is to have this person experiment with modes of behaviour and, *of primary importance*, the experimenter decides what he or she will do. This is the ethical basis of this work.

The facilitator provides primary care for the experimenter and only shares his or her expertize using very gentle suggestions and/or questions, providing the experimenter with an information base on which to make decisions.

The facilitator supports the experimenter in all that he or she may report including fantasies. *Example.* The experimenter may start talking about butterflies in the stomach or a frog in the throat. The significance of such observations belongs to the experimenter only, although the facilitator may be curious and ask elucidatory questions. He aims to increase the knowledge of the experimenter, not necessarily of himself, because if it comes to interest in a butterfly the possible ramifications are way beyond the guessing possibilities of the facilitator – as illustrated in the box. Only the experimenter can know what may happen next.

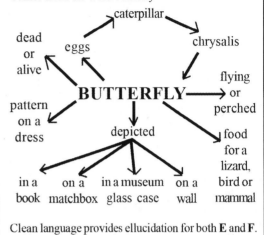

The **E** may use forms of words and the **F** suspects that he or she knows what **E** is refering to. This is not necessarily so as illustrated by a few possible outcomes from the word butterfly –

Clean language provides ellucidation for both **E** and **F**.

Needs, Goals and Ethics

The need for nurture, food and water, is paramount. If needs are not satisfied there can be no goals and no ethics. Ethical values enable evaluation of needs and goals. Goals make the wheels go round.

Wheeler[6] in1991 suggested, following F. Perls[5], that growth is the determinant. That cannot be so because growth depends on the satisfaction of needs for food, water, etc.

The satisfaction of needs, the evaluation of goals and ethics are all expressed in the form of gestalts which come and go, succeeding and supplanting one another in a steady stream of life expression.

Lively human encounters involve two or more people. Each experiments, if healthy, and can be knowledgeable, a facilitator.

Sensations in the body

Feelings, sensations, aches, itches, in the body can be connected to any of many causes or contexts –

- Pain resulting from a lesion in the position of an affected organ or tissue, e.g., an inflamed appendix with the pain in the lower right corner of the belly.
- Pain resulting from a lesion in a position other than that of the affected organ, e.g, an inflamed appendix with the pain apparently on the back. This is the referred pain discovered by Harvey.
- A sensation, often called psychosomatic, that is not connected to a pain source but may give the illusion of so being. Examples are –
  1) gut pains may be accompanied by one or more developed physiological symptoms, for example, diarrhoea, vomiting, sweating, tears.
  2) feeling a heavy lump in the stomach while being very angry about something
  3) having a head ache while feeling out of sorts with

companions. This can provide an excuse for doing something completely different like retiring to bed. 4) A special form of pain occurs when people are in love. The lower gut may be agglutinated with pain when the beloved is absent and the pain vanishes in the presence of the beloved.

It is important to have these options in mind because the first and second main section symptoms need the immediate attention of a physician and/or surgeon. Recognition of the fourth section symptoms may inspire the affected person to write poetry and seek marriage or its modern equivalent.

Felt-what's!

Of particular interest in this essay are fantasy phenomena, referred to above,. The extant symptom may have been identified with feeling-like, fantasy impressions, butterflies or rocks in the stomach, frogs or lumps in the throat, "dark clouds hang over me", the impressions of a wall or other obstruction that seems to appear between self and someone in authority like a bank manager.

Historically these fantasy impressions were clearly recognized by Gendlin[14] in 1981 whose nomenclature – *felt-sense,* has been adopted and adapted. The impressions are now called *felt-protofigures* in conformity with gestalt terminology. Grove[15, 16] in 1989 developed the impressions in metaphor form as a very effective approach to the treatment of catastrophic traumas. In the present gestalt context the impressions are developed through the familiar stages of figure evolution as set out in Chapter 2.

When I acknowledge a felt-sense I know that an *unusual feeling* is occurring within me, usually in my viscera, chest or throat. The felt-sense is not one or more of the following; affect, emotions, thoughts, memories, desires (wants) or needs.

The unusual *felt*-feelings and characters –
- may emerge at quiet or violent times
- occur on the conscious / unconscious border on membranes inside the body
- are whole entities although they may be internally complex
- lack definiteness – unlike emotions, anxiety or excitement which are precise
- change in character, growing step by step,
- are jumpy, sticky, heavy, jittery, tight, etc.
- are difficult to describe and name although a name may be found when an apposite one occurs, e.g., black moth or grey toad
- are such that the presence of a felt-sense may be announced when the experimenter uses a nonverbal noise, uhuh, wow, aha!

The present approach is based on a *hypothesis* –

the protofigure arises from the neocortical part of the protofigure module only, is not connected to the amydala and thus has no affect content.

Justification for this statement will appear in Chapter 10.

Attention to the felt-protofigure can lead to its development to form a complete, new figure and thus a new gestalt, which is related to the initial traumatic event and is free of connection to affect.

In the practical sense the experimenter starts a protofigure session being able to remember a traumatic incident complete with catastrophic affect. He or she ends the session able to remember the incident without the affect. Nightmares, flash backs, motor tics, etc., no longer occur, as Grove showed.

Shadows of the felt-senses occur in ordinary speech and examples are – "I'm liverish." "He's splenetic." "He hasn't the stomach for it!" "He's out of wind." "He's too cocky!" "What an arse hole!" "He should get his teeth into it!" "She's nosey!" "She never puts a foot wrong." "He's a handy man." "The gangster fingered his mate." "He showed his enthusiasm by showing two fingers." "There's a weight on my shoulders."

I have little experience with these shadows. It could be possible to experiment with feeling liverish and explore the development of the concept "liverish" using clean language.

Clean language

Languages are replete with traps for the unwary – ambiguities and verbal formulations that interrupt the flow of communication. Generally the listener has to stop his or her processesing and wonder what the speaker is on about. There is also often good reason to think that the thought processsses of the speaker were in a muddle.

Levitski and Perls[17] selected aspects of language that had a detrimental effect on experimenters during therapy and thus also pointed to beneficial language. Grove[16] extended the selection process and presented a sub-section of language he called "clean language"*. The effect of use of this form of language is to establish calm empathy during which, in my terminology, using the facilitator's "clean questions and statements", the protofigure can be developed through the figure stages until it form a complete, new gestalt.

The gestalt story becomes clearer when certain terminological inexactitudes are eliminated. Principle among these is the attitude to the people involved. The knowledgeable person is not a therapist or counsellor, the all wise leader. He is, to be effective, a gentle facilitator companion, curious about processes and with suggestions for experiments. The person who offers facilitation must be autonomous, self-reliant and capable of choosing on his or her own initiative. The experimenter is suspended in a limbo of experiment, self generated or suggested by the facilitator. The facilitator is a free spirit and models such for the experimenter. The facilitator does, to some extent model for the experimenter by reflecting back what he or she sees or senses.

Circuits and cycles. A second terminological inexactitude occurs with descriptions of the events that occur when the gestalt figure develops. There is a ground, we suppose, and from it emerges the first stage of a figure. After eight stages the gestalt becomes re-grounded, a complete cycle has

---

* A full description of clean, effective language appears in Appendix 1.

occurred. Here is the problem word, cycle. The word infers that further cycles will occur as happens with wheels, biological reproduction cycles, and biochemical energy producing cycles. For the emergence of one gestalt this is not so, we need a word that implies one cycle and that word is circuit; we are interested in gestalt figure circuits.

Neurology. As alluded to above, a modification of vocabulary helps when dealing with neurological events. The founders of the gestalt feast, at the turn from the 19th to 20th century, were clear that a gestalt came from or related to somewhere and, as referred to above, they called it the ground. This was satisfactory for static gestalts as occur in artistic situations. We can now modify our ideas about the ground because we know a lot about neurology[18] and the precursor of the gestalt figure, now termed the protofigure, is activity in the central nervous system. We become aware of minor aspects of the protofigure as felt sensory impressions and these are termed felt-protofigures.

Interrupted and ungrounded figure sequences. In health the figural aspect of the gestalt consists of eight stages developing in sequence. One unhealthy event can occur when the figure is interrupted and there is no return to effect completion. A second style of deleterious event can occur when stages of an apparent figure occur in a circuit but without grounding; this occurs in depression as will be elaborated in Chapter 7. These circuits tend to repeat in true cycling style.

Engagement. The word contact is much misused in the gestalt context. Everybody uses it without thinking. To make contact is to touch and not manipulate. Contact when kissing finds lips touching, that is all. To be engaged in kissing is the next delightful step. Contact in a car is putting the key in the key hole while engagement to fire the engine involves turning the key. Engaging the gears is a step beyond contacting the gear lever. A facilitator and an experimenter can sit together, telephone, skype or stroll in a park and make good contact. To engage in therapy they need to actively, conventionally engage, the one with the other.

Polarities. Behaviours occur in pairs of phenomena. This is illustrated by the NCO who aggressively bullies young entrant soldiers and immediatly becomes supinely passive when the colonel appears. This NCO changed engagement behaviours to suit his situation and in a neurotic way, as S, Freud might have put it. The NCO could benefit by changing behaviour to being assertive while he was still aware of the possible benefits of occasional use of aggression or passivity.

Many other engagement behaviours will be discussed in Chapter 4 where polar contrasting engagement functions will be examined in pairs and the possibilities of resolution evaluated.

The happy worker
has more than one figure in his awareness.
He is throwing his pot on the wheel
He is singing or humming a tune
He is wondering what is for lunch.
He is remembering
that his wife was cooperative last night.

Health

In the scientific medical model lack of health is evaluated in comparison with the healthy state (Appendix 2, page 171). I like that attitude and you will find it throughout this book.

- I have genes that have changed little over the millennia.
- I am part of society in which I learned most of my behaviours.
- I am part of nature, of the living, animate forms that inhabit planet earth. I have two basic needs, to eat health providing food and to procreate so that my line continues far in the future.
- To satisfy my needs I venture out rather like an amoeba. The amoeba bulges out towards food, engulfs the food, then retracts its bulge. I send out hands on arms to grab food, bring it to my mouth, then rest arms and fingers again.

The unit of activity of amoeba or me is a patterned, whole activity which can be repeated as often as necessary. In health all such gestalts are complete units – it is natural for me to seek completion as I am then satisfied and happy with what I am doing and can then get on with next business, a new gestalt. If I feel incomplete, interrupted, disconsolate, I heal myself by completing my gestalts. If hungry, I eat. If tired, I sleep. If randy, I fuck. If I am prevented from completion, hunger, tiredness and/or randyness get in the way of other things I want to attend to.

Generalizing, healthy activity consists in completing otherwise incomplete gestalts.

Engagement Functions

Behind every gestalt is a module* of neuron tissue usually termed the ground and it is clear that in the hurly-burly of gestalts coming and going there must be a neuronal selection mechanisms for the modules so that appropriate gestalts arise at opportune moments.

The gestalt itself is a module of nerve tissue that drives muscular activity, hormonal secretions, etc.

One of the most important ground modules is that for learning – learning about the environment so that nurture needs can be satisfied. This module requires instant absorption of information, etc., and, for maximum efficiency, then requires assimilation for it to become natural for the owner, as described by Anna Freud and F. Perls for what they called introjection. Similar engagement functions will be described in Chapter 5 where due acknowledgement to earlier authors will appear.

---

* Modules are agglomerations of neurons that fulfill functions, e.g., the primary visual cortex in the occipital lobe of the brain is concerned with seeing and the auditory area on the temporal lobe is concerned with hearing. It is evidently opportune to refer to a module, as will occur in Chapter 10, if it has been demonstrated by neuroanatomists to exist.

Declaration

I present in this book many novel aspects of gestalt analysis and need them to be firmly based on gestalt ideas and practices as they have evolved since the beginning of the twentieth century. The early part of the book is thus devoted to what ges erned with field theory, dream analysis or philosophical preoccupations, principally with existentialism and phenomenology, , etc. These for me are ways of avoiding examining what the gestalt system really is so there will be little mention of them in what follows.

Association and dissociation

The coordination of motor and sensory modules is described by neuroanatomists as being by association because neurons link the associated areas. This has been shown[18] for "cognitive behaviour planning, thinking, feeling, perception, speech, learning, memory, emotion and skilled movements." (p 1165). In Kandel's essay lack of association is variously termed deficit, disorder, blocking and damage and can be subsumed under the term dissociation, as the opposite of association, because, for example, in alexia the neurons to the posterior cerebral area do not function.

These are the central pivot points for any gestalt discussion because association associates with introjects and dissociation with projection. These matters will be discussed again in Chapter 4. A consideration of dissociation is important in relationship to failure of grounding described for depressed persons in Chapter 7.

Research and innovation

If you live with a full appetite for life
you only need this book
to facillitate the processes
of someone who is seeking
a full appetite for life.

I have always been a scientist and, in employment, involved in research and making the innovations for the conduct of gestalt therapy reported in Chapters 9 and 10. Those attitudes continued for me as I discovered the psychological world and particularly the gestalt approach though I did not, until recently, think of myself as being a researcher in that field.

The innovations arose out of a knowledge of the basic practices and theories of the gestalt approach and they are described here as such. So the early Chapters are concerned with gestalt theory as it influenced me during my investigations and as I suppose it will influence you when you consider the validity of my observations.

While gestalt figures and grounds are very personal attributes they are meaningless unless in relationship with a similar excitable, energetic and intelligent person with whom to dialogue and dance. Such a person is in the environment (field) which also contributes meaning.

The perpetual search is for holes. What is missing here that I must supply in order to function properly? However, it is all very well to chatter about completing gestalts – what I really want is a square meal and a good night's sleep.

Many holes are represented by protofigures so developing the protofigure to form a full gestalt figure is itself a hole filling game.

House keeping

Abbreviations. For ease of layout the experimenter and the facilitator may be indicated as **E** and **F** respectively.

A felt-protofigure may be alternatively termed a gestalt protofigure and reduced to GPF.

The identity of experimenters has been disguised by use of *nom de plumes* or **Z**, **Y** or **X**. The facilitator is the author unless otherwise designated.

Health and lack of health are discussed in the context of the data set out on page 171.

If you know all about the ideas and practices of the gestalt approach you can start reading at Chapter 6 or 9.

Notes and References

1   Perls, L. (1996). in Smith[21], p 7.
2   Perls, F. (1973). p 74.
3   Wertheimer, M. (1938).
4   Freud, Anna, (1937).
5   Perls, F., Hefferline, R.F. and Goodman, P. (1951).
6   Wheeler, G. (1991).
7   Zinker, J. (1977).
8   Polster, E. and Polster, M. (1973).
9   Philippson, P. (2009).
10 Barber. P., (2006).
11 Snow, C.P. (1966).
12 Kuhn, T. (1962).
13 Popper, N. (1952).
14 Gendlin, E. (1996).
15 Grove, D. and Panzer, B.I. (1989).
16 Grove, D. Personal communication.
17 Levitski, A. and Perls, F., (1951).
18 Kandell, E.R., Shwartz, J.H. and Jessell, T.M. (2000).
19 Darwin, C. (1859).
20 Buss, D.M. (2008).
21 Smith, E.W.L. (1996).

The butterfly
counts not months
but moments,
and has time enough.

Rabindranath Tagore.

—=(☆)=—

Part 1

Classical Gestalt Processes in Health

When asked: "What are the goals of gestalt therapy as you see them", Laura Perls replied: "Ongoing Gestalt formation. What is of greatest interest to individuals … advances into the foreground where it can be clearly experienced and dealt with. Once resolved these interests can move into the background which then leaves the foreground free again for the next challenge – the next gestalt … . Closure is always temporary. There is always movement onwards". (in Smith, 1996, p. 25)

Men, women and children who live in a world of peril are compelled to seek safety. Two ways are available – find or build shelters and forts and devise defensive weapons or create mythological, theological thought systems that deny the existence of danger and/or promise rescue. The person needing therapy has lost part of both aspects of the art of self-nurture – not just for shelter and food, also for the satisfaction of other needs that are necessary for a full and healthy life. Many people spontaneously relearn the lost arts or may relearn them in cooperation with another person, a Shaman or, in our times, Grandmother or a counsellor/therapist/facilitator.

In what follows I will refer frequently to self-nurture, the dynamic state of self-actualization and self-realization, and I want to make clear that, whilst I am interested in what each person does or does not do for him or herself. I am also interested in what the person does in co-operative nurturing and association with other people. This happens all the time in childhood if the child is to survive and thrive and is equally important in adulthood, in the family, work-place and in play.

To state these points another way, I am also concerned with self-wounding, how people do it and how people heal themselves. Thus whilst my key personal concerns are with nurture, memory and learning I am aware that Laura Perls also wrote –

We have an obligation to see that Gestalt Therapy does not become a fixed gestalt.

(in Smith, 1996, p. 25).

Chapter 1
Gestalt Figures in the Environment

Musing on my life style and my life space I have produced an algorithm (Chart 1·1) that summarises my ideas and feelings, linking in the needs I must satisfy in order to stay alive aong people, what I perceive in my environment that can satisfy my needs and an indication of the action I must take in order to obtain satisfaction.

Derived from these concepts is the recognition that there are three main forms of gestalts to be considered –

a theory of perception and
a theory of action to satisfy needs

These gestalts will now be considered in detail in the three main subdivisions, satisfaction of needs, perception and action It is important to make these distinctions because entrapment in ideas about perception is a passive, static state, like mentally sick people, whilst action is assertive and dynamic like thriving, lusty, healthy people.

I find that I am, in setting out these ideas, putting gestalt ideas in an evolutionary context, being my need for adequate nurture and my need to father children.

Historically perception was considered first so it is appropriate to do the same here.

Chart 1·1. WE MEET AGAIN

Tuning in; the "chemistry" of relationships

Hello Friend!

Well met!
I am excited by you.

We gladly associate

We keep off intruders
as we extrude towards
one another

I enjoy my
affinity with you

Contact is followed
by engagement

The perfect tight bond =
two women talking about shoes –
difficult to break in.

In a good conversation

We associate and engage

I put out a valence
towards you and

You put out a valence
towards me

We bond
I am entranced
by you

You are entranced
by me

Multivalenced chat =
easy to break in

There is nothing else
on your mind – only me

There is nothing else
on my mind – only you

Women are said by
women to be good at
multitasking

I associate with you

I do not dissociate from you

Another tight bond =
business folk talking about a contract –
very difficult to break in.

THE GESTALT THEORY OF PERCEPTION

Various authors defined gestalt in various ways. The poet Goethe (1749 – 1832) said a gestalt was "the self-actualizing wholeness of organic forms." Dilthey[1,2]: "It is through awareness that [one] comes to appreciate the whole in terms of its parts." Ehrenfels: "gestalt qualities are inherent in a phenomenon and can be discovered there. The total form is something different from the sum of the elements." Wertheimer:[3] "The structured whole, or gestalt, is the primary unit of mental life." Ash:[4] "Gestalt means form in relationship to a whole something[5]." The

> The essential differences between
> figure and ground
> are that the figure
> has form qualities,
> while the ground has none,
> and that the figure appears
> to have "thing" character,
> while the ground has
> only the quality of
> undifferentiated material.
> Rubin,[5] 1914

form of the entity, the gestalt, that is observed depends on the nature of object and ***the mind-set of the observer.*** In forming definitions I find it necessary to be pedantic and to be clear about these connotations, while I also differentiate between two forms of gestalt, static and dynamic.

Interpretation and fantasy

Köhler[6], as an example, experimented with many volunteers assessing their perception of various static patterns. There was little reaction to uniform, homogeneous fields of vision but with inhomogeneity the people described what they saw in varied ways. Köhler interpreted the data from his observations more extensively than is relevant to the present argument.

Lewin[7] emphasized that the reality of perception, thought, will and emotion has elements of "wholes of experience,"

Chart 1·2. What do you see?

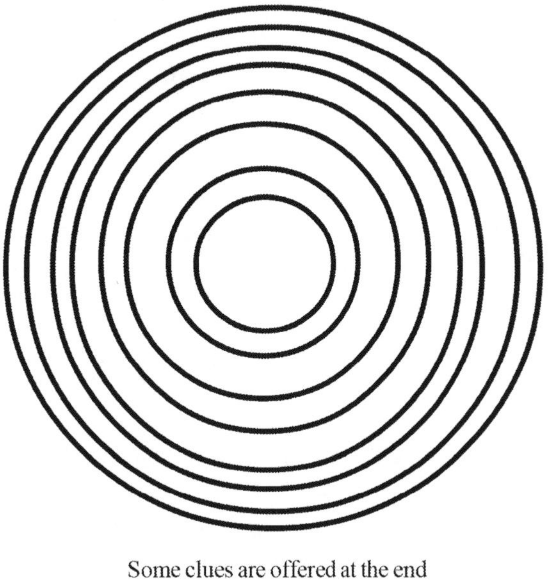

Some clues are offered at the end
of the Notes and References. Page 31.

and understanding elements (parts) is derived from the character of the whole, the gestalt, in which they occur. This is "a procedure which leads at once to a consideration of the outer and inner fields in which behaviour takes place." In this context Lewin considers "temporally extended wholes," referring to extension in time, with "continued repetition or direction towards a goal" thus extending gestalt theory contemplation of puzzles (here illustrated by Chart 1·1 and also see Rubin's two heads / vase drawing[5]) to events in real time. Lewin" was also interested in the role of energy in mental happenings. "In any instance of a psychical [mental] process we must inquire after the origin of its actualizing energy." "Perception can … directly arouse new intentions or desires and … energy capable of doing work is set free." "Behaviour is steered by the perceptual field." The concept of a goal; "With the attainment of the goal there are comes [a] … change in environmental forces. Satiation involves not only a change in locus of these forces but also a decided change in the psychical tension which had underlain the goal-seeking behaviour." Such events as "emotions, intentions, wishes and hopes are embedded in specific psychical units, personality spheres, and behavioural wholes," gestalts. The interdependence of minor gestalts depends on whether they are or are not part of the same major gestalt in a hierarchy. Thus "satisfaction of one desire will more or less carry with it the satisfaction of some other desire." But this may not be so. "Unity of consciousness," writes Lewin[8] is not evident as the self, as a gestalt, "is only one complex functional part within this psychical totality." "We may say that psychological unity is that of a weak gestalt comprising a number of 'strong gestalten'." the latter are only "in part … in communication with each other, in part disclose no unity at all." Each gestalt, however, exerts its tension. Pairs or more of these may be in opposition resulting in attainment of an equilibrium position. Thus I may need to urinate but be in a place where this is socially and hygienically not possible. Resolution of these counter-tensions occurs when I reached a toilet and discharged both tension systems. Lewin discusses the roles of time, intention, purpose, opportunity and expectation on the mode of consummation.

Schulte, like Goldstein, observed patients and introduced a more dynamic approach, when describing how paranoid patients became as they were by adjusting to their circumstances and then needed further adjustment when being "cured."

Adjustment is also interpretation. Gestalts are not absolute entities – ***we interpret and make sense of what we see and hear.*** Concentric rings (Chart 1·2) are seen and confusion occurs when interpreting. What do I see as I hear Beethoven's Pastoral Symphony?

Wertheimer[3] in his book, on "Productive Thinking" codified his ideas about the nature of gestalts, the perceptive and active forms as I described them, and I have summarised his criteria in tabular form (Table 1·1). Many of the concepts of Köhler and Lewin are subsumed in Wertheimer's ideas and I will concentrate on the latter. Wertheimer did not make the distinction between perceptive (static) and active (dynamic) gestalts and, as I see it, his exposition thereby lost clarity. When it comes to gestalt therapy, the distinction is vital because healthy people are, as noted above, active (dynamic) in attitude, whereas unhealthy people tend towards stasis. Before pursuing the clinical path it is useful to exemplify some aspects of what Wertheimer said about thinking, which are an important contribution to self-nurture.

Thinking and seeing

Wertheimer[3] in his essay on the dynamics and logic of *Productive Thinking* describes his methods but only reticently entitles them as a gestalt approach some 15 pages in. I must admit to finding Wertheimer's generalizations to be set out in a somewhat awkward manner and found that his account became more accessible when I held in mind that he was writing about perceptive and dynamic gestalts[9] without clearly differentiating between them. I gained clarity when I used concrete examples when thinking about his propositions. Such as –

- A mandala, an example is shown in Chart 1·3 – is it complete and providing artistic satisfaction, is something missing, do the parts bond together?
- A SATB choir is singing Bach. Is one tenor singing sharp in his enthusiasm or has an alto turned over two pages?
- A group of mechanics is assembling an aeroplane and discovers that the store man has supplied the wrong sized wheels.
- An old fashioned pendulum clock is telling the wrong time. What can have gone wrong with it?

The reader will need to refer to Wertheimer's book to assess whether I have represented his arguments accurately and adequately. However, Wertheimer[10] says that all healthy people have an appetite for real, productive thinking. Reassembling his propositions somewhat and retaining his emphasis on the thought processes, problem solving (this is also Köhler's emphasis) and the general appetite of people to get to the core of a matter we thus have a gestalt approach which includes operation of traditional inductive and deductive approaches together with a holistic attitude which goes beyond them in the ways set out in Table 1·1.

The table illustrates many points of gestalt theory. They characterize whole gestalts based on Wertheimer's observations. Each has sub-gestalts set off with a strong vertical interaction along the left-hand column, with weaker interactions cross-wise. In thinking about inner relationships I experimented with many alterations to Wertheimer's original sequence, particularly when I found a gap in the internal logic. Thus at one time I separated items concerned with thinking from other items. When I was concerned with what Wertheimer meant by the words "sensible" and "improve" I separated my concern off as a footnote.

A story: two players assemble 96 dominoes in a game played by rules. Though only one person wins both have the enjoyment of creating a unique pattern of pieces on the table.

These aspects of gestalt analysis are worthy of more detailed examination, considering the occurrence of thoughts, interruptions of development of thought processes, the differentiation of the figure from its ground, the differentiation of figures from gestalts and the role of the field in which everything exists.

Thoughts as whole phenomena

> When I complete a jigsaw puzzle
> I can enjoy the created picture

One of Wertheimer's concerns was with thoughts about the gestalt approach to thinking, seeking whole-phenomena, groups of units, and recognizing a centre. Is a gestalt formed or forming? Is it weak or strong? Does it involve me? He[11] saw the emergence of gestalt theory as the conceptual solution to the logical problems of science and philosophy of his time.

Wertheimer's suggestion is to perceive each process item as "in varying degrees as structured (gestaltet), each consists of more or less structured wholes and whole-processes with the whole-properties and laws, characteristic of whole-tendencies and a whole-determination of parts." Severance of a part from the organized whole in which it occurs involves alterations in that part. "Modifications of a part frequently involve changes elsewhere in the whole itself."

Chart 1·3 An imperfect mandala
dedicated to Alice
who found her way through a looking glass.

He[12] gives as example: "the cells of the organism are *parts* of the whole and excitations occurring in them are to be viewed as part-processes functionally related to whole-processes of the entire organism. [This is the way the physiologist approaches the matter]. Likewise we react to something exciting: not to mere sensations.

He[13] then asked a pertinent, crucial question. "What happens .... when one suddenly 'sees the point'? particularly when a problem becomes solved? A process has occurred during a period of contemplation, often quiet, is interrupted by an abrupt realisation. In fact the adventitious event is a whole unit, a concept, an idea, something to do, a gestalt. Like any other gestalt it has structure and may be part of another structure, a larger gestalt."

Wertheimer examined scientific method and saw thoughts as breaking phenomena into parts, and then re-assembling them. Gestalt theory, however, is concerned with "wholes, the behaviour of which is not determined by that of the individual elements, but where the part-processes are themselves determined by the intrinsic, inherent, nature of the whole." This, for him, is the fundamental "formula" of gestalt theory. This "is not only an *outcome* but a device: not only a theory *about* results and observations but a means towards further discoveries" [something inherently native: the concrete process]. "See the dynamic, functional relationship to the whole" ... in order to understand.

The field, the environment of the figure and the gestalt, Wertheimer says, encompasses me. [He used the word ego]. Again the laws of whole-process apply – I'm a part-process within my field as a whole process. The field is not a summation of sense data. The field is a complex of interlocking, interacting, processes. A man in a community is such.

Similarly psychology requires an approach from above and not from "below upwards." Wertheimer goes on about concentration on stimulus rather than on sensation and on the elucidation of mind-body unity: but this is not, for me, a clear exposition. He makes too many diversions without clear examples.

Problem solving by a group of many people requires "centring" – for success each person gets into a role equal to the others without one-upmanship or domination. Each person takes as much time as is necessary to do what the person and the group want the person to do, a process often called brain-storming. In this way emotional processes help the intellectual processes. The group members can ask themselves: "where is the centre for me," while expecting something serendipitous.

When contemplating real world gestalts, in contrast to mental constructs, constituent parts of gestalt theory can be detected. Some of these in summary, culled from Wertheimer, are as follows.

Wholeness. The human tendency is to always finish whatever is going on. Satisfaction follows completion.

Change. Stressed in the points above, is change. A gestalt only ossifies if it is a static figure like a mandala. In the healthy personal life and similarly in the life of groups of people, dynamic change from one equilibrium position to another is inevitable. Self-organisation and self-reorganisa-

---

Gestalt concepts in concise, abstract form

These concepts are utterly basic to the gestalt approach and are worth reinforcing in an interim summary.

I may be aware of a multitude of phenomena, each of importance roughly equal to the others; this is my *environment* or *field* of awareness. I become aware that one phenomenon is becoming more interesting to me and eventually becomes of greater interest than any other phenomenon. This gestalt *figure,* when fully emerged from the neurone complex and mature, occupies my *foreground* and is a complete entity standing out starkly from the *background* of my environment. My figure, once its function is complete, recedes in importance to me and effectively reverts to neurological *ground* and I remember it. A new circuit of figure formation may then begin for me – if I am confining myself to an interest in one figure at a time. My figure may be weak, strong or of medium strength and has, hypothetically, a boundary as one of the features that differentiates it from the ground or any other figure.

On a beach by the sea, among hundreds of naked people, a man becomes aware that his penis is showing his sexual excitement. For him his erect penis is figure standing out from a ground of the rest of his body. Then he becomes concerned about himself and what people are thinking about him and he, himself, becomes figure standing out among the rest of the people.

A marigold seed falls to earth and, in due course of season, begins to grow. The roots in contact with the earth take up nutriment and eventually flowers appear followed by seeds which fall to earth. The plant dies and rots and nurtures new growth of other plants. Just so a healthy figure maintains its roots to ground and when dying back into ground provides nurture and a model for subsequent healthy figures.

---

tion is the criterion of health. Variation is as intrinsic in gestalt theory as in Darwinian theory.

Simplicity and prägnanz. Gestalt writers frequently refer to prägnanz. Köhler says this occurs when "the best possible shape is perceived, a perfect circle or a square." This may be generalised to the statement that any gestalt will be perceived or actually exist in the simplest form possible. Prägnanz of the gestalt is the tendency towards the simplest shape with, if necessary, a reorganisation of components and diminution of energy, towards the "simplest and most regular grouping." Köhler:[14] "the tendency [towards] the simplest shape" or form. When some gestalt writers refer to "the minimum principle" they are referring to prägnanz. This is Ockham's simplicity idea.

Process and equilibrium. An organism at rest has its systems in a state of equilibrium. Excitation causes a

Table 1·1. Characteristics and qualities of static gestalts

Based on Wertheimer's[16] observations and criteria. My additions are indicated with *.

Characteristics of static gestalts. They –	Synonyms, etc.	Characteristics of non-gestalts
are **Natural**	Gestalts are real.	
are **Observed**	Noticed as units; perceived, known, be cognizant of, recognized, discovered, discerned.	Not noticeable as a prominent, outstanding unit.
are **Expressive**	Meaningful, an emanation, originative, arising.	Minimally expressive.
are **Complete**	Whole, an entity, a unit, an entirety, a. totality.	Have a sense of incompleteness.
have **Form**	Structure, pattern, conformation, configuration, style, arrangement, organization, design.	Little or no significant structure or pattern.
are **Holistic**	The ready recognition of whole items rather than parts.	Bits and pieces.
have **Symmetry**	Harmony, balance, consonance, conformity, congruence, regularity. A patterned gestalt will show **symmetry** in the disposition of its parts.	Little or no sign of symmetry, etc.
are **Simple**	Prägnance, lucid, clear, plain, orderly.	Are complicated; muddled.
have **Similarity** of sub-units	Seen likenesses, resemblances, affinities, kinship and homogeneity. The smaller units will probably be very **similar** to one another. [similar molecules coalesce into crystals – the chemist's method of purification of a substance].	Smaller units that might have agglomerated into a gestalt were too dissimilar so that an intact, integral pattern does not form.
are **Grouped**	Sub-units of a gestalt, being similar, tend to come into proximity, become coalescent, ordered, close and form a continuous pattern. [The ancient Greek key pattern]. Appearance of a core, pivot or focus. The process of knowing is often a process of **grouping** to form an orderly, whole unit by structuring the parts and sensing a centre for them.	Disparate parts may line up but do not constitute a whole. [Like a series of random numbers] Disparate parts have no tendency to group or coalesce into structured units.
are **Centred**	Sub-Groups in a gestalt often appear to have a **Centre.**	Formless, unrelated bits and pieces rarely show signs of having centre.
are **Separate**	Distinct, independent, disconnected, bounded. disengaged from surroundings and enclosures.	There is no demarcation among bits and pieces.
are **Emotionally satisfying**	A pleasing pattern.	A jarring agglomeration.
are **Intellectually Satisfying**	A sensible, reasonable and intelligent pattern.	No sense of reason.

movement to a new state of equilibrium. This appears to be "purposive" and "goal activity" [perhaps goal orientated] towards organic differentiation and growth. "Causality" is thus "directed by the total pressure of the system." When a goal is implied Wertheimer refers to a "purposive gestalt."

Closure. Koffka[15] describes closure as the tendency to recognize patterned effects, closure or recognition of closed spaces rather than open ones. The perfect gestalt is closed "proceeding towards the intrinsically appropriate end of a behaviour sequence." If a jig-saw puzzle is assembled and one piece is missing the player searches diligently to enable completion.

Similarity. Wertheimer[16] discusses the tendency of like parts to band together, to unify, like sheep in a cold field,

Proximity. Köhler: "every interruption of a physical steady state produces a rapid process of displacement which can then result in a new steady state. [I guess that his steady state is a state of equilibrium].

Problems. Wertheimer says that problems always start somewhere and there is an end in view, no matter how vague

the initial concept. The problem is the beginning of a figure and the solution, occurring in the particular present, often requires maturation time and/or occurs as a serendipitous, satorian, flash impression "Ah Ha!"

In addition to the general observations concerning gestalt perception, Wertheimer and Köhler showed that two other phenomena were of importance during the development of the figure –

1) there was a threshold below which perception did not occur and above which it did.

2) there was a point of development at which something completely different occurred. Thus a person may spend some time solving a problem, then the solution, completion of the gestalt, comes suddenly. This is change of quantity into quality or an increase of quantity that is suddenly interrupted by development of something qualitatively different from anything that has gone before.

Polster and Polster[17] comment on the contrast of speed of movement, smoothly from figure to figure for OK people and with jerkiness for people with problems.

THE GESTALT THEORY OF NEEDS

A reminder: in this chapter I am intending to deal with core self in the most perfect healthy state – though actually I cannot do so because, as a member of the *homo sapiens* species, I am imperfect and living in an imperfect, largely hostile environment. However, I can do my best towards being perfect.

The first thing I must do is deal with my needs. Maslow[18] taught me to be aware of a hierarchy of needs so I set them out in a special box called *Needs and Appetites* and I can give them attention from time to time.

How do I become aware of my needs? I recognize internal

signals, appetites, that tell me about needs. I usually do not choose to give attention to appetites – they come up on me serendipitously and then I give them attention.

Do my needs relate to my modern, culturally determined state or are my genes impinging stone age determinants on me? (Page 139).

What happens to me as I settle down to deal with my needs? I become concentrated, all my faculties become devoted to this one task. I find that I am fascinated and have set out some of the characteristics of fascination in a box of that name. In this state of fascination I can do anything.

Needs and Appetites

For staying alive.
For breathing.
For excretion.
For curiosity.
For children running free.
For a lover.
For sex and pre-sexual activities,
including groping.
For progeny
For progeny's progeny.
For a pet cat, dog, canary, etc.
For peace and quiet,
For the land, horticulture and gardening
For good literature – a play, a sonnet.
For graphic arts – except for modern formless daubs
For scientific investigations and experiments.
For exploring, country walking and geography.
For adventure.

I am Fascinated

My figure is strong.
I am totally interested in what I am doing.
Every part of me is doing what I am doing.
I have no boundary with what I am doing.
I fully engage with what I am doing.
I am involved with one purpose only.
I am not curious about anything else.
I am communicating from my metaphorical heart.
I am not aware of anything else.
I effectively have no background.
I have no sense of internal conflicts.
I am concentrating
but not aware of concentrating.
I am generally, gently pleased with myself.
I am hardly aware of time passing –
time may even seem to stop.
I am happening.
I am involved in organismic self-regulation.

When I am fascinated and excited with someone energy develops until a sudden change occurs – an emotional outburst, laughter, usually, but sometimes anger or weeping.

In healthy activity each emergent self-nurturing figure proceeds to completion and, as activity effectively ceases, satisfaction is felt, all is remembered and effectively reverts to ground. Affect is expressed, ideas are known, engagement occurs with another person, activity is accomplished and the healthy person goes on to be fascinated with new figures in an effortless flow.

When fascinated I am not concentrating with clenched teeth, I am not involved in Reich's moralistic regulation[19] or Perls' deliberate regulation." "… the object occupies the foreground without any effort, the rest of the world disappears, time and surroundings cease to exist: no internal conflict or protest against the concentration arises." F. Perls[20].

Tom Brown bowled the cricket ball towards the opposite wicket. As he ran, his right arm circled under and backwards and swung over his head. His hand released the ball at the apposite point and continued swinging until the movement was completed. The wicket fell.

A mother bringing her small baby to her breast epitomises all that can be said about healthy interface; the baby drinks and there is no rigid boundary. Lovers have intact physical boundaries and physiological boundaries, no matter how the amorous struggle transforms them, and are only penetrated by passing sperm while their psychological boundaries are practically non-existent.

Wertheimer wrote about completion of gestalts as appetites and discussed the appetite for solving problems.[21] Other appetites, without trying to be comprehensive, are set out in the *Needs and Appetites* box.

A further observation of Wertheimer is very interesting. "Children possess an original urge to find out things which produce as much satisfaction as sucking or being stroked. The task of education, then, is to support the child until it arrives where it basically wants to go." Likewise, the task of the facilitator is to facilitate the processes of the experimenter as he goes where he liminally or subliminally wants to go.

THE GESTALT THEORY OF ACTION

Everything is in flux. Only after we have been stunned by the infinite diversity of processes constituting the Universe can we understand the importance of the organizing principle that creates order from chaos; namely, the figure ↔ background formation. Whatever is the organism's foremost need makes reality appear as it does. It makes such objects stand out as figures which correspond to diverse needs. It evokes our interest, attention, cathexis or whatever you choose to call it. F. Perls[20].

Returning to my main theme of self-nurture as the basis for a person to survive and thrive it is clear that all material things brought from the environment into self involve movement of matter in space and time, work is done (as in physical mechanics) in activity, and changes to self and environment occur. And having recognised two types of gestalt, static and dynamic, it is a short move to the realisation that any personal gestalt is dynamic and extended in time, whether in micro-seconds or months. A figure develops and forms its conformation, as Wertheimer emphasises (Table 1·2), which implies that in observing and supporting figure development, the facilitator is aware of passage of time. This time interval feels like *now* and indeed it is so providing that such a *now* is recognised as extended in time. Likewise the place of action, the here of the incident, has vicinity as Köhler emphasised. The facilitator is thus interested in what the person is doing in an increment of time and place rather than in an instantaneous "here and now." This more highly personalised attitude was discussed by Polster and Polster[17] and remains a definitive attitude for gestalt facilitators. This corresponds with the view of Smuts[23] who considered only the dynamic aspect of evolving organisms, particularly of self-regulation and self-balance.

> Ask a busy person
> if you want something done.
> Old Kentish saying.

Of course in other contexts F. Perls also emphasized his primary interest in what the person was doing. If he over emphasized the "here and now" attitude[24] he was probably expressing the insecurity that beset him when concerned with past and future.

The active person is functioning to some purpose with, to emphasize again, the usual mode of action being recognition of and dealing with needs, etc., and gaining satisfaction. In departure from normality the following divergences in relation to purpose and need can be recognized –

- Loss of function   – underdone activity
- Gain of function   – overdone activity
- Change of function – avoidance of organismic
  intention by doing something
  completely different. These
  may be seen to be undesired
  side effects and the products
  of loss or gain of function.

Two figures running simultaneously, or nearly so, can result in confusion – which figures shall have main attention and be acted on? – or the figures may have contradictory outcomes. Tom Brown, while bowling the cricket ball, suddenly thought of the danger of injuring the boy who stood in front of the wicket. Tom's hand released the ball before the apposite point was reached and his feet stumbled. The ball went for six, the crowd roared and Tom wanted the opportunity to bowl again, to show that he really could bowl well. Here Tom simultaneously wanted and did not want to project the ball in the direction of the wicket and the batsman.

Table 1·2. Characteristics and qualities of dynamic gestalts

Based on Table 1·1 and on Wertheimer's[16] observations and criteria.
My interpolations other than examples, are indicated in square brackets [ ].

Characteristics of dynamic gestalts. They –	Synonyms, etc.	Characteristics of non-gestalts
are **Natural**	Gestalts are real.	
are **Emergent**	The tendency to begin, be a source and develop [Often accompanied by a need, appetite, etc.] A whole gestalt **emerges** from a ground. Wertheimer says it "springs forth" from the as "not yet formed."	Disparate bits and pieces only make a rubbish dump.
[are **Targeted**	Development with intention to satisfy a need, appetite, curiosity, etc.	Rarely develop in a purposive manner].
are **Expressive**	Develops with meaning, as an emanation, with original, arising characteristics. [With awareness and attention]	May be vaguely expressive
are **Complete**	"Entire of itself." Develops towards a whole, an entity, a unit, an entirety, a totality. [A folk dance is an unique entity proceeding to completion].	Rarely show signs of completion.
have **Form**	Developing with structure, pattern, conformation, style, configuration, arrangement, organization and design.	The potential gestalt failed to form because of an interruption process.
	During formation of a gestalt, smaller . units come together, these may themselves be gestalts, and **form a pattern.**	A crystal may form with dirt inclusions
are **Holistic**	Readily recognized as whole items rather than parts. For Wertheimer the **whole** is often recognized before one is aware of the presence of parts. [Like a sleek car].	Parts may attract immediate attention. [Like a crashed car].
have **Symmetry**	Develops with harmony, balance, consonance, conformity, congruence and regularity.	Rarely shows symmetry, etc. Incongruent. Dissonant. Complex.
are **Simple**	Develops with lucidity, clarity, plainness, and orderliness. Depending on circumstances, the product gestalt tends to be as **simple** as possible, "good" as Köhler and Wertheimer write, and this is the principle of prägnanz.	Simplicity is lost as complexity disrupts the gestalt formation.
have **Similarity** of sub-units	Agglomeration, unit upon unit, with likeness, resemblance, affinity, kinship, homogeneity. [Photographers tend to join camera clubs].	Rarely show much similarity.
are **Grouped**	Develops coalescence in an ordered, close manner with proximity of parts.	Rarely develops in orderly groups

Table 1·2 continued.

Are **bounded**	Have clear, distinct surfaces and borders, differentiated from the environment,	It may not be clear that there is a distinct surface or border.
are **Centred**	Appearance of a core, pivot or focus [Often a need satisfying action. Amateur actors come together to perform a play which becomes their shared centre of attention].	Rarely show a centre.
are **Separate**	Develops distinctively, independently, disconnected and disengaged from. surroundings and enclosures. [Performers of **Hamlet** occupy a different stage from performers of **Lear**].	Tend to merge boundaries
are **Dependent**	What is perceived depends on the observer; a person interested in motor bikes will tend to see them all over the place.	The unaware person notices little.
are **Emotionally Satisfying**	Develops in a demonstrative, fervent, ardent, passionate manner, specially when other people are involved.	Uninteresting, boring.
are **Intellectually Satisfying**	Develops in a sensible, reasonable and. intelligent way.	Puzzling, perplexing.
[are **Retrogressive**	Develops to the point of natural recession, and regression. All good things come to an end].	May seem to never end.
[Have **no encumbrances**	The figure develops without unwanted material	Extraneous baggage detracts from the effect of the main subject].

While I follow Perls in being interested in spontaneous activity there are other aspects of life that are also exciting. Activity can be automatic and yet not be spontaneous. I decide to throw a ball but I do not set out to remember every muscular exertion necessary to do so.

There is also calculated activity as occurs when brain and hand on paper solve an algebraic problem.

Self-nurturing, self-responsibility, cooperation and self-support

An absolutely healthy person
Is completely in touch
With himself and reality
F. Perls[24]

The fully healthy person gets on with life, flowing into all the activities that nurture self. This is self-support, as F.

Perls calls it, and is also taking responsibility for these activities. The only modification to this pattern of activities is in cooperative relationship with other people. I can suggest that I will look after the drinks and prepare the sweet course while you prepare the main dish My emphasis on cooperation is a major departure from F. Perls' mode of action which was always to emphasize individualistic characteristics, a rather pugnacious American style, although it seems that he willingly co-operated in writing with other authors.

Denial of self-responsibility occurs when a person exploits his or her neighbours instead of making suggestions with clear statements, manipulation is used. The polarity here relates whingeing and pleading with authoratative, aggressive demands, Each is an alienating condition for decent, cooperative people and usually relies on hooking the attention of some other alienated person.

The therapist is not operating from alien-self and does not support attempts to force responsibility on himself or on any other person. The exception is when violence against others or self-harm threatens. "If we are to continue to work together I want a contract with you that if you think of suicide (violence, self-harm, etc.) you will also think of me – telephone or come – look after yourself."

If the client tries to transfer responsibility, block the move, keep the responsibility with the client. "I can't cope!" "*In what way can't you cope?*" Acknowledge the state of the client and go for elucidation. "I need your help!" "*What will happen to you if I refuse your request?*" said with no reference to needs, wants, etc.

The verbal indicator of shared responsibility is the use of the generalized "you," rather than the shared "we" and other forms of dissociated language: "such busy times." It is standard gestalt practice to suggest a change, the first two to "I"s and the second to something like: "I am always busy." A more effective form of repost is: "*I notice you said 'you' – if you mean me that is not true.*" Similarly for "we" – "if you include me that is not true." As for busy times: "Who is so busy?"

Self-nurturing and problem solving are total life stories, are a sort of intellectual glue that holds all life together and provides consistency. The person comes to therapy because the glue has failed, pieces no longer adhere. Part of the task in therapy is to help the client sort out the pieces and fix them together again – with the proviso that this process continues for the rest of life. (Here I declare a debt to Eric Berne for his neat explanation of life-scripts). It is not so much the births, marriages and deaths that are of interests in therapy but where the client starts his story and where he stops.

The story is a schema of *snakes and ladders*, alien and progressive adventures. The story as told is, of course what the client wants the therapist to hear. The skill is in finding what the client wants, by decision or otherwise, to keep secret.

It follows that F. Perls' statement[26]; "[G]rowing maturation is the transformation of environmental support to self-support" is partially erroneous. Cooperative, mutual environmental nurture is also valid. F Perls' statement merely reflects his own egotistical isolationist attitude.

The clearest way to differentiate "neurotic" dependence from social dependence is to consider the self-nurturing aspect which is directly related to need satisfaction. Cooperation between equal people is OK: exploitation is not OK.

Just so is the cooperation of the client and the therapist: one who needs knowledge and one who has it. And the process is one of facilitation, of experiment so that the partners remain equal. The story is a process of processes.[27]

Pleasure is OK. I can take as much as arrives. Pain is a

Chart 1·3. The figure's pair of boundaries.

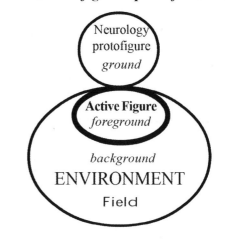

warning that something is wrong – maybe my leg is broken, or this is a psychic warning of a received insult. I hate her and I can't bring myself to say so. If this pain is a fantasy it is rubbish and none the less effective. Anyway, I avoid pain, get my bone mended or my mind trouble sorted out – otherwise I go phobic, F. Perls says so. However, he also said "unpleasantness … invariably changes into pleasantness" if stayed with. So don't ever run away – stay at the impasse. If I "interrupt [my] awareness continuum as soon as [I] meet … something unpleasant, I then run for cover and support, withdraw, invent excuses (mind fucking says Perls), role play (poor me), play Bernian games, become melodramatic, abusive, depressed in order to avoid the simple contact with the unpleasant." Would that it was so simple. It becomes habitual to avoid the unpleasant, it occurs out of awareness, it seems to be the normal thing to do. While it is not noticed by the player of these games and external observer, Granny or a therapist, recognizes what is going on and shares awareness.

The human personality, excitement and growth, these were the stated sub-texts for Perls, Hefferline and Goodman when they assembled their account of *Gestalt Therapy* published in 1951. And these are my main interests when I enjoy looking at myself and the people in my village or therapy room. Does my excitement spring forth at appropriate moments giving rise to genuine emotion[28] or do I stifle it and prevent the growth of my personality? Is my sense of my own humanity waxing or waning? When I talk to a neighbour, does his excitement burst forth or do I see it quenched?

The Encapsulation of Figures

Each active gestalt figure has two boundaries as illustrated in Chart 1·3. The obvious boundary is with the environment in which it exists. Thus, on page 15, Tom Brown was operating in a figure to do with throwing a ball and he had a border with his environment which included a batsman and a wicket. He also had a boundary in his mind/brain, among his neurons, because he remembered previous experiences of

projecting a ball on a cricket pitch.

This latter is a neurological module, in evolutionary psychological terminology, and for convenience in this gestalt context it is termed a protofigure[29], the neurological module that is destined to form the gestalt figure in the environment[30] where the activity occurs. (See the footnote on page 5).

The neurological boundary

It has probably been obvious from the inception of gestalt ideas that, for perceiving living organisms, a figure emerges from a matrix of neurons in the mind/brain. It would have been assumed then, that the nature of this matrix would never be known. However, Gendlin noticed what he called felt-senses emerging into awareness and centred in strange parts of the body, mostly in the guts.

Edwards[31] recognized that these felt-senses were early aspects of figure formation and called them felt-protofigures. An example makes this matter clear: **Duddy** presented the facilitator with a story about his right hand. Sometimes when he was writing the hand became tremulous so that he could not write. This lasted only a few seconds. Having told his story he became comfortable and relaxed.

F   As you are quietly wondering what happens next you could notice an unusual impression somewhere in your body.

E   [After a short pause] I have a tremor in my stomach.

This tremor is the felt-sense the felt-protofigure. As Gendlin showed, it can best be discovered in a trancy, meditation state. By concentrating on the qualities of the felt-protofigure it can be developed to form a full figure because said felt-protofigure is stage <1> of a new gestalt figure. The matter of the permeability of the neurological boundary is taken up again on the next page.

The environmental boundary

Latner[32] has a delightful analogy – the boundary is where the sea churns on the beach. Lewin[33] stressed that equilibrium of tension systems presupposes a functional "boundary and factual separation over against its surroundings."

The meaning given to a concept is modified by the nature of the context. Thus in –

$$\begin{matrix} F \\ 10 \times O \cdot 1 = 1 \\ U \\ R \end{matrix} \quad \text{and} \quad \begin{matrix} E \\ Y = mX + c \\ T \\ R \\ A \end{matrix}$$

the meaning of the characters "O" and "X" depend on the embedding figure, as the line is read horizontally or vertically.

A corollary of the figure $\leftrightarrow$ ground theory is that the figure must have an hypothetical *border* in order to be differentiable from the context. Thus "mX" and "EXTRA" each effectively have fantasy lines round them to include their space and exclude extraneous information, the contiguous field.

I find that there is a problem here. If I consider a boundary I am beset with fences, well demarked division positions. And I find that this is not so for healthy people. An active person has little impression of any separation between himself and what he is doing; there is flawless free flow. I will return to question boundaries again in Chapter 3 where the topic will be limitations to activities in unhealthy states. Meanwhile I call my interface with my world my border. If I was stuck it would be a boundary, implying a holly hedge. Just as water has a surface, so I have an interface with my world. To pursue this metaphor; just as water is held in a vessel by gravity so do I retain myself intact within my surface. An hypothetical pull holds me in – except when I want to venture out or take nutrition in.

Lewin[34] pointed out, everyone lives in an hypothetical capsule, in an irregularly shaped box, enclosed to some degree, packaged and surrounded by a barrier used to keep in what is self and keep out what is extraneous other. For some of us the capsule is hardly present and effectively fully permeable, a simple border. For other people permeability may be limited. The shy person has a host of feelings and

---

There, here, next

A particular figure emerges from a particular ground, into a particular context at a particular time, that time is the *now* of the circumstance and the place is the *here* of that circumstance. F. Perls expressed one of his interests as: "I maintain that all therapy that has to be done can only be done in the now. Anything else is interfering."[35] Perls emphasized repeatedly that the relevant state for being clear about the figure $\leftrightarrow$ ground relationship was, at **that** time and place, to be *here*, *now*. Thus in the *Four Lectures* he[22] says:

I have one aim only; to impart a fraction of the meaning of the word *now*. To me, nothing exists except the now. Now = experience = awareness = reality. The past is no more and the future not yet. Only the *now* exists.

The curative figure always grows in a particular *now* which has the possibly infinitesimal time increment of that figure, and only that figure. However, if I burnt my finger in the past I can know what a hot object will do and avoid burning my finger *now*. If I become short of milk and potatoes I need to plan a time in the future when I will buy what I need so that at some future *now* I can prepare a meal for my benefit.

So I acknowledge a restriction on Perls' edict. In the therapy room work is best done in that *here* and *now*. Anything from the past or expected in the future is brought into the therapy room and acted out using present tense language. In everyday life it is not beneficial to continually be "here and now." For one thing it would antagonize companions. The past provides the anecdotes that fuel conversation and the future provides the basis for mutual exploration of options.

reasons why other people are unapproachable and must not be allowed to approach. The bombastic person wonders why people shrink away. The capsule may be hypothetical but the slogans and excuses that hold it in place are innumerable beliefs about self and others that hold the personal world within bounds and sustain a sense of security, holding off vulnerability.

The nature of the three dimensional frontier that is the capsule has been extensively considered and some of the observations are summarized in Chart 1·4 where the contents, point by point, serve to emphasize the rich, enormously interesting, variety of conditions in health. The paucity, uniformity, negativity and alienation in lack of health will be taken up in Chapters 3 to 5).

I find that there is much advantage in paying attention to environment in which I see the experimenter to exist and calling it the field as does Lewin. My context marks me out from everything that is not me and everything that is me illustrates me, my personality, to myself and any observer.

The interpersonal interface

> We touch, we get in contact, we stretch our
> boundaries out to the thing in question
>   F. Perls[36]

Polster & Polster,[37] following F. Perls, describe a hypothetical boundary for the self as the place of engagement of self with the external world and describe it as the I-boundary. Similarly, Swanson[38] who was concerned with morality[19] provides an essay which is a rich source of ideas about the personal engagement boundary. This boundary is hypothetical and is not contiguous with the skin, the aura or any other physical or not physical entity.

> The contact boundary is the point at which
> the organism and its environment meet; it is here
> that dangers are rejected, obstacles overcome [and]
> assimilation occurs.

These and other observations of Swanson, summarized in Chart 1·4, are is marked with numbers {X} to correspond with the following discussion.

The engagement of the organism with its environment forms the basic polarity of life {1}. Whatever is within my interface is natural for me {2}, I am responsible for whatever it is {3}, I like it and I may love it {4}. Whatever is organismically me defines my morality issues; it is beneficial and right for me. The intimate, personal nature of my organismic self emphasises my unity, wholeness and uniqueness. And a neat emphasis on these aspects is provided by contemplating the polar opposite, what is not me and outside of me. A fully real person can cross his or her interface with full self-support and without fear. In a group the facilitator and experimenter were working together. Another group member interrupted in an empathic and facilitating way. Later the facilitator gave the latter an accolade for crossing into what might well have been a dangerous space, for example, if the facilitator had rejected the interruption.

Whatever I deal with inside myself forms my view of what is real in the outside world {5} because I use subjective processes to create my assertion of objectivity about what is not mine. In doing so I correct and deal with the external world and discriminate to be sure that what I take in is what I need or want {6}. In the process of taking in I make mine of whatever I take in, I mark it with my personality. Discrimination is an important aspect of my active relationship with my environment. I go out {7} when working with other people and I withdraw when that suits me. I thus keep my personal life as clearly my own while also being able to maintain an active social life. Discrimination also provides my mechanism for dealing with problems {8} because the outcome lies, for me, in recognizing what is me, what is mine, what I want and need for myself, and reject the rest, just as I do in a dangerous situation {9}. My recurrent theme is encapsulated in the question: "How do I deal with foreign matter from my environment?" {10}. I reject, alienate and separate off what I don't want. I accept, incorporate, acquire, embody, adapt, adopt, assimilate and merge into myself what I have evaluated as mine. In doing so I form a new unity for myself; myself of a little while ago augmented by the new matter of experience, making a new self. Any such approach to my milieu is likely to be risky and I determine on the particular moment, at the particular place, what I will do next. Sometimes a risk {11} is potentially overwhelming and I withdraw. Sometimes my fears around the risk are negligible and I advance. Provided that I can be fluid in my approach, rather than fixed, I can cope. More than coping, I can thrive.

Permeability of the neurological interface: Protofigures

It has long been accepted that what goes on in the mind/brain is not accessible unless it becomes a proper conscious process, a gestalt in the present terminology, and it will do so without much volition, or if hypnosis is used.

Vague fantasy feelings have been known, presumably since ancient times, including that there are "butterflies in my stomach", "ants crawling on my skin", "black clouds hanging over me" and lots more. Gendlin[39] (1994) showed that these "felt-senses" could be utilized to form the basis of therapeutic manouvers by focusing attention on them. He had some difficulties, however, because he had to teach his "patients" meditation techniques for some weeks before applying therapeutic techniques.

Grove and Panzer[40] (1999) obtained direct contact for therapeutic purposes by using a subset of language that they termed "clean language". (A description of clean language appears here in Appendix 1). Grove considered that his process revealed metaphors that could be subject of attention by way of developing the character of the metaphor.

Edwards[41] (2010) showed that Gendlin's felt-senses and Groves metaphors were aspects of gestalt figures. The

Chart 1·4. Healthy events on my interface with my environment.
This analysis is based on the observations of Swanson[38]

A = assimilation processes

The discussion in the text refers to the sections designated {X}

INTERFACE

	Organism = Me = Self	**Environment = Not me = Others**
{1}		
{2}	natural for me ← **A**	unnatural for me
	I am responsible for myself	You are responsible for yourself
	good for me, right for me,	wrong for me, bad for me
	like, love	indifferent, hate, dislike
{3}		
{4}	personal, internal	external, field, milieu
	associated, O.K., known	unknown, not O.K., dissociated
	identification, native	foriegn,
	attraction, inalienable	alienation, repulsion
	inherant	strange, estranged
	ingrained, intrinsic	extrinsic, extraneous
	intimate, familiar, essence	separated, unfamilair, dispassionate
	innate	severed, disjoined
{5}	introspective, subjective	objective
{6}	I need, I want ←— **discrimination** —→ I reject	
{7}	private ———→ **sharing, helping, collaborating** ——→ public, communal,	
	personal ←———— **withdrawal** ←——— cooperative,	
		participatory
{8}	becomes part of me ←— **dealt with** ←— obstacles and problems	
		danger
{9}	**dealt with**, rejected	
{10}	incorporation ←— **assimilation** ←— foreign	
	aquisition **evaluation**	
	embodied	rejected
	adapted, adopted	alienated
	united, merged	separated
{11}	taken ←— **risk** —→ rejected	

psychological terminology, and the grounded precursor of the figure was termed protofigure to be consistent with gestalt terminology. The aspect of the protofigure that became conscious (butterflies, ants, clouds) was termed the felt-protofigure to be consistent with Gendlin's and gestalt terminology[39].

Most importantly it was shown that the felt-protofigure, as an aspect of the gestalt ground and part of a developing figure which could, by use of clean language, complete the development of the figure and thus of the gestalt. This is the therapeutic act. However, this anticipates the matter of chapters 3 and 9 so further discussion will occur there.

Permeability of the environmental interface

Having formed the hypothesis that there is a personal interface between self and others and the environment we are obliged to make further hypotheses. The first is that in health the interface has the property of being both discriminatory and permeable on a voluntary basis – it lets in only what is wanted and lets out everything that is not wanted. The hypothesis extends to lack of health when the boundary lets in unwanted things and does not let out what is not wanted.

The second hypothesis is that taking care of self occurs on the interface as alluded to in the first hypothesis. The third hypothesis follows from the second and is, as Perls and colleagues say, that the boundary *is* the self. The fourth hypothesis follows again and is that, because of the discriminatory function of the interface, attention, awareness and action occur there.

Self-support and social support

All goes well as a personal figure crosses the interface into interpersonal or interenvironmental action if accompanied by adequate courage, if the venturer supports self-nurturing rather than sabotaging self. However, in a given situation it may be beneficial self-support to withdraw since personal safety is of basic importance. Otherwise the challenge of the situation may be met, the risk taken and the adventure ventured.

Frustration occurs if a recognized need is instantly denied – "I have no right to feel randy right now" – if energy development is cut off – "I'm not going to bother to go to the kitchen" – if engagement is avoided – "I could talk to this interesting looking person but his cockney voice is awful" – if action is avoided – "The eggs are ready to cook in the pan but I don't know where the matches are to light the gas."

Self-support is provided on two levels. Most importantly by running the basic life-support figures of need; for O_2 in, for CO_2 out, for food, for drink, for excretion. self support also entails cuddling up to the partner, having energy rise, searching for food when hungry, accepting harsh accents of voice when engagement is wanted, and actually boiling the eggs for 4 minutes to make them edible.

These are all aspects of engagement and action and, as L. Perls wrote:[42]

> Contact can only be as good as the support that is available ... Support is everything that one has assimilated and integrated. What has not been really integrated ... becomes a block, becomes a fixed gestalt which is in the way of ongoing gestalt formation."

Once in the foreground the block, the unfinished

situation, can be dealt with, finished. Self-support takes many forms. An aggressive person invades the space of other people who, having their borders violated, take defensive action or withdraw. A passive person does not approach his border directly although he may manipulate an aggressive person to make the border crossing. The assertive person gently approaches his borders – "Are you there? I would enjoy talking to you if you are interested" and is ready to engage and join action with another assertive person.

In childhood parental support is essential. As adult character develops a wider social support is necessary as the person works for a living and probably marries and produces children. He or she also becomes part of the wider matrix of social support when assisting others . Problem solving needs to be neither a lonesome task nor require brain-storming group activity. It is what friends are for.

Attention and awareness

Attention and awareness are most importantly concerned with what is outside self. They are also the medium of appreciation of inside phenomena from bellyaches to fantasies. They are also my mode of existence as I expand my experience and knowledge. Attention and awareness are manifestations of the permeability of the personal interface for figure emergence. Concentration, likewise, can be concerned with anything, inside or outside. Interruption, obstruction, resistance, however, occur on the interface in the form of disruption of figure development. The agglomeration of behaviours, seen as a whole, is described as personality or character. Personal development requires attention to events on the environmental interface, correcting inadequacies and reinforcing beneficial behaviours. There is much here that needs elaborating, so here goes. Part of my problem, as I write, is that I am now concerned with many words of closely related meaning; adding to those used above, knowledgeablity, knowing, nous and gumption.

- What is the focus of my attention, right now?
- The usual question among gestalt practitioners is: what am I (are you) aware of now?
- Am I aware of my awareness?
- As I am aware and attentive. Am I concentrating?
- When I am with someone, what is the focus of his or her attention, on that moment?
- My attention is on what I am aware of, the subject of my current figure.
- Your attention is on what you are aware of, the subject of your current figure.

From our mutual attention stems our relationship. Close attention provides a close relationship. A wandering attention by one or both partners leads to a poor relationship. If I am attending to something, that something is the centre of my awareness, the subject of my current figure.

Attention is not necessarily a continuous phenomenon. One aspect is that attention lasts only as long as interest is maintained, an aspect of figure life. Another aspect is that interruptions occur and these merit very serious consideration because badly managed interruptions are self-defeating and lead to a deleterious disposition. Attention to events is synonymous with awareness of the event. Interruptions are found to be satisfactory if followed by return to the original event.

In the therapy situation the experimenter is the centre my focus of attention and I am interested in his/her focus of attention. Since the experimenter has a therapeutic engagement with the facilitator the expected primary centre of external awareness for the experimenter is the facilitator. "If a ... client isolates himself, then I might invite him to have contact with me." (Marcus[43]) Awareness of emotional expression is of prime importance: "The awareness of, ability to endure, unwanted emotions" are the "essential conditions for a successful cure." (F. Perls[24]). And the most important among these emotions are being terribly sorry for having done something wrong, guilt, shame, disgust and embarrassment.

Sharp presence is easier under the sun, under a clear blue sky, with a light, warm breeze off the sea. The challenge is to maintain sharp presence if the sky is grey, clouded, cold perhaps and with a wind to cut to my bones. I hear, see and feel sharp: sounds penetrate my awareness, I see and I can attend to my meanings – she is talking, is it to me, what is she saying? – my reply flows as I hear myself. Am I gruff or easy? She wants something and I won't give. I see full colours, sharp definition, precise boundaries. My shirt clings to my sweaty chest and the skin of my back tingles.

The awareness continuum is not a list of sounds, sights, etc., an inventory. It is not sensations. It *is* an expression of experience (Naranjo[44]). I like the word *continuum*, it goes on for ever and a day. It leads to productive experiments if experience is contrasted with a game with lists. I remember the Zen statement that to give something a name is to make it vanish; thus I was fascinated by the colony of a hundred beautiful birds on the hillside in Palmitos Park, Gran Canaria. When I gave them their name they vanished and were replaced by memories of bedraggled, lonely peacocks seen in English parks. I struggled back to reality and saw again the gorgeous display before me.

The Self

My concept of the nature of the self makes self an active entity rather than passive. Nevertheless every person is aware that self often does things on autopilot so that one has a concept of underlying, unchanging propensities. These as a whole are usually called character or personality.[103] This concept has always been regarded as useful – novelists put us in touch with their characters by describing character. The concept need not be foreign to gestalt theory because the whole character differs from the sum of its parts in the usual gestalt manner. Character analysis is valuable provided that the holistic context is always in view.

The core self can thus be described and in a sense analysed. Because dynamics and stasis are so important it is necessary to contrast nurturing and deleterious processes. Core self, as L. Perls says, has style (page 1), a neat expression of the unity of the healthy person who is not beset by anxieties, fears, guilt or other self-defeating behaviours and ideas.

The core-centred person, spontaneously expressing him- or herself comes into conflict with other people, most importantly his or her carer adults, parents, teachers, etc. This is necessary. For a child to become an effective, fully socialized adult the child must learn to read, to write, to count, to know the history of its forebears and stand on her or his own feet. The child learns about risks and safety or it will not survive

Since this chapter is concerned with the healthy forms of gestalt psychology it is necessary to postpone further discussion of character and personality to Chapter 4 which is concerned with self-wounding and introduces a fuller discussion of matters alien to civilized people.

INTERRUPTIONS OF ATTENTION

The person who continually discounts
the abilities and activities
of other people
wonders why, he would surely say,
each other person, so many of them,
becomes angry.

An interruption is disruption of a particular aspect of attention. It is a negating process which may occur for external reasons, for example, the telephone rings, someone arrives or an earthquake occurs. Or an interruption may be internal: "I've got a head ache coming on so I don't want to talk to you." Both forms of interruption are self-expressive and to a degree self-nurturing.

Lewin[34] described tension systems where resolution is not immediately possible resulting in a widespread energy expression as with a child who throws a tantrum. Lack of completion of the gestalt being communicated results in bad communication because of what is lacking.

If an interruption occurs, as written above, all is well provided that there is a return to the initial situation which ultimately becomes finished.

Lack of specificity during communications

Lack of spontaneity is a form of interruption because an important aspect of evidence of what is being talked about is missing. Levitski and Perls[115] wrote about the **unspecified subject**, not that they called it that. Here *it, that, this,* and *they* are used instead of the subject in such a way that the listener has no idea what the subject is and is manipulated to ask a question towards elucidation. The meaning of the *it*, etc., can be clear if it is near enough in the sentence to the subject matter. Otherwise clarity is gained by repeating the subject. It is, of course, one of the games that couples play when one expects that the other will remember a previous subject, sometimes stated over an hour or more before. This effectively envisions that the other can thought-read. A source of much agro'.

A commonly used **verb** of poor specificity is *to do*. Which leads to the question, "Do what?"

Commonly used **adjectives** of vague specificity are *coloured* – "Which colour?"; *weight* – "what weight?"; and many more.

Unspecified signifiers may also cause confusion if the designation is not clear. "Go to the house" – "Which house?" "Whose house?"

Truth and falsehood. When a person interrupts his discourse and says something like: "To tell you the truth …" he infers that everything else he says is false. Watch a politician on the T.V. if you want to see this ploy at work as he or she tries to emphasize veracity.

Misleading statements. When someone says: "I'm afraid …" it usually means "Sorry to say …". "Sorry to tell you …" and has no emotional content. "Don't you see …" is usually about ideas and not about something visual.

Chart 1·5. Embedded interruptions from everyday life.

In this chart the rising lines represent the rising limbs of the figures, the horizontal lines represent the action stages of the figures and the descending lines represent the completion of the action stages followed by completion of the figures.

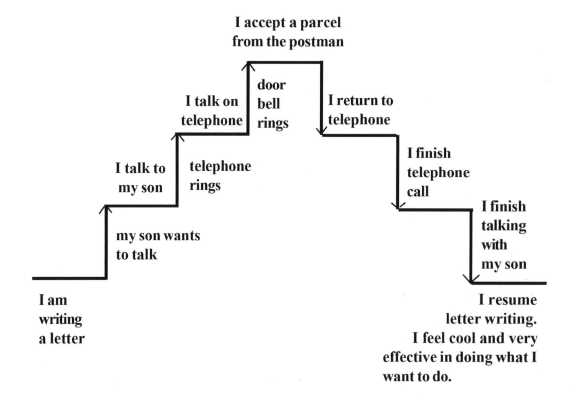

Embedded interruptions

I write a few sentences and then go to the kitchen to turn my sausages that are cooking under the grill. I return to write a few more sentences and go to the kitchen again for a mug of apple and cinnamon tisane. Provided that I return to where I left off I remain happy. I hop from one incomplete figure to another and eventually all figures reach completion. I finish writing a section and then enjoy sausages and tisane for supper. This type of situation is illustrated more elaborately in Chart 1·5.

I write a few sentences of this page and then think of something to add in the introduction so I go there. After attending to the introduction and returning to this page I get a strong idea about depression and add it to chapter 8. I don't need to care about interruptions provided that I return to, add to, and eventually complete, what ever was interrupted. Alex, on being interrupted says: Let's put this on the back burner for a while, and deal with your problem."

That is enough for now about interruptions, beneficial and lethal. The latter is a subject to return in Chapter 3.

Engagement and withdrawal

Engagement with self-nurturing ways of satisfying needs is essential for life and in this context are a multiplicity of forms of engagement including with other people, with food and drink and with clean air. However, all forms of engagement are intermittent. The period of breathing in oxygen is followed by breathing out CO_2. Eating, in our times, occurs three times each day. I come together with other people in occasional co-operative feeding endeavours. Termination of life occurs if mutual engagement is absent or overwhelming, for example if continuous drinking results in neglect of essential nourishing activities. So successful living results from a balance of engagements with the many needs for the sustenance of life. Departure from balance is followed by return to balance in trouble free life and this may occur automatically or with the aid of deliberate awareness of what is going on.

Effective engagement is characterized by focus of attention, concentration on satisfaction of needs with adequate energy and affect – not too little so as to require pushing and not too much so as to overdo things. Thus, inadequate or frustrated engagement and activity are characterized by alternating or simultaneous and confusing incidence of under- or over-energization.

Another characteristic of effective engagement is often described as "the experience of differences." This is an interpersonal effect. Sometimes people agree to co-operate with one another and sometimes they don't – it is essential for each to be able to say "yes" or "no" freely with feeling and ideas.

Some writers have considered only the interpersonal aspects of engagement and this is important, of course. Simkin[45] says he sees a need to clarify engagement (he writes about contact) into the forms of interpersonal, interenvironmental and intrapsychic and these terms are self-explanatory. For him the last is evidently the state of "knowing" that interpersonal contact has occurred without the usual engagement functions, conversation, touch, etc., indicating a Healer's energy exchange. As Simkin emphasizes, the opposite of engagement is withdrawal. Among essays towards engagement one may become aware of a hierarchy of needs to be satisfied including those of both people and the general environment. One person prefers to eat without conversation while another needs to chatter between mouthfuls. Yet another likes music in the background.

The next issue to consider would be the mechanisms for the deleterious breaking of engagement but this will be dealt with early in Chapter 3.

English language usage is based on $I \leftrightarrow You$ relationships; it is judged to be archaic if anyone uses the $I \leftrightarrow Thou$ relationship. Nevertheless, we can take the intimacy message in Buber's sense, in contrast to the $I \leftrightarrow It$ relationship where the other person is treated as an object. This has been much discussed in the gestalt context of dialogic relationships, particularly by Yontef.[46] This overemphasis on dialogue was exposed by Rogers who pointed out that empathic relationships often involved no dialogue at all as happens with the holding hands and adoring gazes of limerant lovers. The effective facilitator loves the experimenter with an all encompassing gaze (no staring), while holding hands is all too rare. The experimenter quietly gains confidence and, when ready, talks . In Amsterdam I met an Austrian man of approximately my own age. We largely communicated silently in bed and when we spoke, used a mixture of German and English. Part of our closeness was expressed by my following his suggestion of using thee and thou language.

Buber concentrated on the power of intimate relationships between people, expressed by them in knowing self as *I* and the other as *thou*, with the grammatical variants *thy* and *thee*. This is very neatly expressed by Schoen[47] in his essay on Buber. Thus a special I-attitude of mind is always potentially ready to relate to the other person as *Thou*. If A is relating to B it is all very well if A enjoys intimacy with B in an $I \rightarrow Thou$ state – if B does not respond, and may even become hostile, the relationship does not work. The ideal may be represented by –

But ideals are often far away. **Abe** was a rather softee man and approaches a woman he does not know in a friendly way. He doesn't know that she is a feminist who reacts aggressively to all approaches by men, reckoning that she alone can decide if engagement is OK. So **Abe** gets hurt.

There are, of course, as Buber established, ways of relating that are successful. In England I can use *You* as if it was *Thou* or *Thee* and *Your* as if they were *Thy*. It is only by auxiliary words, tone of voice and gentle gestures that I can achieve any indication of *Thou*-ness and then I assess myself as coming from a calm centre. A poorly engaged self would produce a rough tone of voice for *You*'s and over done or absent gestures.

The comfortable limits of memory and awareness

Zeigarnik[48] studied people in a laboratory setting and made many valuable observations summarized by him as –

> Unfinished tasks are remembered approximately twice as well as completed ones … the recall-value of unfinished tasks is high because at the time of the report there still exists an unsatisfied quasi-need.
> The quasi-need corresponds to a state of tension whose expression may be seen … in desire to finish the interrupted work… if the task has not been completed to *the subjects own satisfaction* (Zeigarnik's italics).

Zeigarnik was a colleague of Lewin. The "quasi-need" is clearly a reference to what we now call an unfinished gestalt. The explanation was in terms of conflicting "tension systems" occurring when the director of the experiment interrupted the experiment. The experimenter wanted to continue the experiment and also wanted to obey the director. This polarity conflict energy attached to the unfinished task, ensuring that it be well remembered. There was no such energy attached to completed tasks. They also stated that the recall-value of interrupted tasks is highest in ambitious subjects and that both fatigue and generalised excitement impair recall.

Perhaps I think an event is all over. Soon I know it is not over since I remember varying amounts of detail and can recount them to myself or another person. Memory is an important function of ground and fortunately or otherwise can become contaminated by invention, myth, excuses, suppositions and "if only …"s Such developments come from the tendency to turn what is remembered into a fully blown figure, embellish it and interpret.

Here, then, with memory is the direct link of extinct figures with extant ground, modifying the ground and thus modifying the development of future figures.

Man, of half a million years ago, needed to remember important things for survival, where the deer graze, that a wolf lurks and kills, where home is, as examples. In our times we are bombarded with irrelevant information, mostly provided by power-hungry rich people who want to become richer, mediated by the advertising system. Whereas our ancestors forgot at their peril, we, in the early 21st century, have much that we must forget to survive. My awareness must be restricted in relationship to my needs. The individualism of late capitalist society has two pungent aspects of character. I and my needs are paramount for me and I am constantly told that I must conform and consume. This dilemma produces opted out, alienated individuals who cut themselves off from social contact to protect themselves and then largely hide their thoughts. In their state of alienation and social dissociation they often describe themselves as depressed.

Self and society

The "[G]estalt – holistic view is that people work best towards common goals rather than in opposition." (Wertheimer[49]). Also "man is not only part of a field, but a part and member of his group." The group of people have a natural tendency to cooperate towards completion of their mutual gestalts.

When such a group of people coalesce some are more active than others. After a time of mutual interaction the energy level tends to a mean, highly energized people calm down and sleepy ones gain impetus. One way to express this state of affairs is to describe the energy as tending towards an equilibrium state. Such an energy state is of itself not constant, there are variations about a mean energy.

Self is a social being and an emulator. Here are some examples –

- A friend of mine who ran a stall in a street market in London would ask his friends to stand by the stall looking at his wares. The fact that some people were looking attracted other people to see what was going on.
- There was a time when everybody would go to the village fête.
- The monkey parade occurs at some point in the village as the teenagers get together wondering what to do.
- When Diana, Princess of Wales, died nearly every British citizen found a place to put flowers.

These phenomenon of gregariousness are rather like hysteria; though gentle, non-frantic and quiet events.

Self also tends to be conspicuous like the butterfly which must be noticeable so that it and its mate may meet and procreate. It must also not be discernible so that another creature will not find it and eat it. I need to be out, conspicuously in my world to satisfy my needs. But I must beware! – a more powerful person may entrap me and use my abilities for his ends. In clearer words he will employ and manage me.

The extant figure, if satisfying needs, wants, etc., is of benefit to the person concerned – that is the only possible criterion. If interruption of process occurs, the figure is not completed and an urge towards completion remains. The action function has not happened. Alternatively non-organismic activities will have happened with concomitant failure to satisfy a real figure. These aspects of engaged and alienated function will be taken up again in later Chapters, after a more detailed consideration of other aspects of gestalt theory. As emphasized at the outset, evidence to support these gestalt and dialectical ideas will be presented from experience with clients in groups and students in psychology class.

Protofigures in the gestalt ground

Efflorescent gestalt figures gain easy attention but what is happening in the ground? To elucidate using an example – as I write my most recent gestalt had me in the kitchen preparing a mug of coffee. In the early form of this figure a feeling in my mouth lead me to think about drinking.

Before that, in much more remote pasts, I have made coffee, drunk it and felt satisfied. I thus have memory ideas about coffee and what it does for me.

Generalizing, it seems that the practical coffee figure was preceded by events in my ground (mind, central nervous system). This precursor event was largely out of my awareness and none-the-less real. As a forerunner of a figure I shall call it a protofigure. The gestalt equation becomes –

Dynamic   = protofigure   +   active gestalt figure
gestalt   in the ground   in the environment.

These protofigures are probably akin to Ziegarnik's quasi-needs. The nature of protofigures requires further elucidation but this I hold off until Chapter 9 when I will have dealt with –

* how gestalt figures develop, stage by stage,
* the deleterious effects of interruption of gestalt figure development and
* the observations concerning depressed people who each had a personal circuit of behaviour that was irrelevant to a healthy life style.

The Zone

I live more livingly
after watching live Actors
cavorting in a theatre

Perls was, by many accounts, an aggressive man, as said above, and he exaggerated freely so I am not surprised that I am unable to accept all his ideas. I can accept his dictat on the value of a decision to continue and complete prevailing figures which is also a decision to abandon the barren stuff that hangs around the ground. The general idea is to encourage the client to reproduce in supportive company what he wants to change, and thus move into self-support. He can demonstrate his interruptions to himself. Then he can make decisions.

Such awareness of events occurring on the instant becomes interrupted and attention is diverted – it is one of the skills of living to be able to survive interruptions and thrive. This may involve flowing with events or making a conscious decisions to stay with fascination for a particular event. Then attention can be given to personal interfaces or boundaries and whether they are to stay intact or be allowed to be open. Figures can then serve needs. Memory can provide information. Contradictory interests can arise and be dealt with. Risks can be taken and curiosity satisfied.

An athlete friend of mine described for me an illuminating way of viewing much of what has been set out above concerning core self. It is life in the **zone**. "There", he said, I am present in what I am doing with no intrusions and distractions, from family, job or other people. I am peaceful, calm, tranquil, positive and only able to make correct decisions. I have no mind other than focus on the single thought. Here I will win my race. I am only vaguely aware of everything going on around me on that moment. I am courageous and confident."

He said he got into the zone by finding his inner calm state by surveying the state of his body, feet, legs and all the way up to his head. "How goes my breathing?" If there is time he focuses on a koan.

How wonderful, I thought as he spoke, that the state of mind for peak performance of body and mind is so important for athletes that they give it a special name. The zone is, for me, the α-state of meditation. It is where I am naturally me, where there are no problems. What more healthy state of body-mind can there be?

He was so tough
they wrote on his gravestone:
"Who're you looking at!"
Overheard in the *Pilot Boat* pub.

GESTALT FIGURES IN CONFLICT

"In two minds, I am"

The banal ↔ profound relationship.
A basic polarity

When discussing repetitive phenomena F. Perls[50] said:
One of the polarities is always hidden . . . we become aware of the inefficient underdog part in ourselves, but we are not aware of the character of the top part in ourselves. Our righteous behaviour, we take that for granted. And the balance between the submissive behaviour and the bullying behaviour, between aggression and the frightened, cannot be achieved.

F. Perls exaggerates as usual; both poles may be strickingly evident. A healthy person frequently has doubts about

doing something. One figure can be out front and the other nagging in the background. Otherwise two figures can openly exist simultaneously in foreground and be of contrary effect. Then a conflict of interest occurs showing as confusion. Such pairs are said to be polar opposites. Perls had a famous client who said he loved his Mother while his foot kicked out.

Splitting of the personality was much discussed by Klienian therapists (Solomon[51]) usually in terms of a healthy aspect of a person who wants to change versus a self-destructive, omnipotent aspect.

Perls and colleagues[52] described many polar opposites, emphasizing the role of a hypothetical "zero point," a neutral or indifferent place between the poles.

Polarity phenomena

In my experience living a single figure is a rare event. Dialecticians usually state baldly that phenomena emerge new born as a pair of polar-opposites. In psychology this is probably mainly true except that, as in the Perls quotation immediately above, one of the poles may not be directly observable. It may take time to seek out the polarization effects and this is particularly important in therapy where a person dislikes the polar position he or she finds self in.

I have discussed dipolar situations above and now add that complications occur when strong figures of mutually exclusive effect occur –

- Simple contradictory interests, dilemmas, as in F. Perls' example[53] of a singer in the middle of a performing choir who becomes desperate to urinate.
- Interests are in direct conflict as with a decision made to visit a friend at 7 pm *and* wanting to watch a TV programme at 7 pm (no video recorder available).

The state of being "in two minds about something" and being torn for choice between two possibilities is quite common. Anther way of expressing being in two minds is to say "I am on the horns of a dilemma." Most people can solve the problems of such a state of existence. For others it can he very confusing.

As far I remember there is no discussion of polarity in the essays of Köhler, Koffka or others among the originators of gestalt theory, in an artificial, experimental situation. It seems impossible that there should be such since the emergence of a figure *is* emergence, for them, of one figure and not two.

This I define as a polarity of the first kind where there is an apparently intimate relationship of two possible views, tails or heads, with no logical or emotional connection. More realistic would be the event in which I am talking to someone. I am interested in the development of our conversation, he is figure for me and *while* I am in concerned engagement with him I cannot conceive of an opposite figure.

Certain verbal indicators can alert one to the presence and activities of dipoles -

- *Negativity* ia automatically a polarity issue because the negative cannot be processed mentally until the positive aspect has been appreciated.
- *But, however, yet* and other prevaricating words and phrases introduce disparate ideas. I want to walk in the park but Ballykissangel is on T.V.

My action polarity is –

Involvement ↔ Withdrawal

If I am aware of what is going on I have a choice among two factors. Again a polarity of what I call the first kind. There is no middle place between these poles as with ground and figure. A polarity of the second kind which has an intermediary concept is exemplified by –

Aggressive ↔ Assertive ↔ Passive
character   character   character

a clear and well established example. Here only the

assertive person is fully authentic: the other people indulge in self-defeating activities, as discussed by Perls[54].

Awaiting the outcome of a process is not necessarily passivity. Patience is essential in many circumstances where disturbance of the equilibrium might destroy the chance of continuing homeostasis. Patience is accompanied by continuation of the poised interest, the sparkle in the eyes and other signs of potency. Passivity is a dead state with little sign of interest in anything. Similarly high energy assertiveness is not aggression. The latter goes with ill feeling, domineering, poor evaluation and lack of consideration for others, all evident in the relevant situation.

As a first move, the state of assertiveness may be gained by avoiding the states of passivity and aggressiveness. When two self-assertive people come into conflict neither hurts the other and debate ensues happily and continues, maybe with some time breaks, until both people win.

When I am aggressive I bellow and shout
When I am passive I speak softly and quickly
fearing that nobody will be listening
When I am assertive I speak steadily,
confidently and carefully

F. Perls[55] in "Ego, Hunger and Aggression" placed great emphasis on the role of aggression and, as he saw it, its polar opposite defence. The latter was seen to be particularly disadvantageous if it turned into the fixed state and associated with Reich-style armour. Perls' "defence" can be seen as either/or what I see as aggression or passivity. Perls also saw the situation in general terms of polarities. "Between each pair of opposites stands the zero point, the equilibrium position, the pivot about which movement occurs towards one pole or the other."[56] This polarity of the second kind, pivoting about equilibrium, covers the whole organism and the simplest way of making an assessment " is to ask the experimenter to convey . . . whatever he experiences mentally, emotionally and *physically*." (Perls emphasis). Perls' zero point sounds like a static state. For me this is not so. The intermediary position is as fluctuating and dynamic as the polar positions. It is perhaps a brevity, exaggeration and an awkwardness in Perls' language that obscures his recognition of the total dynamics of the polarity situation. Perls position is –

Manifestation 1 ↔ Zero point ↔ Manifestation 2
the opposite of   the pivot   the opposite of
Manifestation 2   the balance   Manifestation 1
                             point

So dynamically, dialectically related phenomena tend to return to a central, equilibrium position. For me the zero point can usually be named in a polarity of the second kind.

Another key polarity of the second kind occurs when considering the relationship of the past with the future in the form of the resolution of the concept of *now* –

Before → Now → Next

These relationships are time dependent. Now I can remember what happened before; now I can fantasize about what happens next. Now abouts I can only be here abouts, except in fantasy –

Now — Here

This is not a polarity. It is an expression of the time – dimension – consciousness unity. Likewise, for Perls, the concept of *zones* –

Inside ↔ On the boundary ↔ Outside

a linear dimensional dependency with time involved minimally during movement.

The problem in discussing polarities lies in the tendency to see only static, black and white, colourless, representations due to the limited nature of language. The head side of a coin exists immutably fixed to its verso, the tail. And yet there is metal and an edge between these two. My initial tendency was to use words related to *verso* and say, for example, dynamic versus static. If I do so I immediately imply a combative state (C.O.E.D definition of versus) and I prefer assertion to aggression. The Flight Sergeant who bullies the erks unmercifully is humbly obsequious when facing a commissioned officer.

There is more reality in considering a continuum, polarities of the second kind with intermediate stages, for example –

cold	↔	warm	↔	hot
uncomfortable		comfortable		uncomfortable

and/or –

hypo-activity	↔	assertive activity	↔	hyper-activity
"depressed"		happy		"manic"

How depressed or how manic is happiness, is akin to how hot or cold is warm. A particular place on the scale is in a mobile state and certainly not static. Thus it is highly stimulating to swim in a cold sea and very relaxing to lie in a hot sauna. The mood of a "manic / depressive" experimenter oscillates about on such a scale.

When discussing good and bad F. Perls[57] stated that evaluation, appraisement, ethical systems and morals are important to people but are not aspects of figure and ground relationships. For him "moralism" upsets proper evaluation which can be effected by organismic means. This is evasion by Perls: all his criteria are part of the processes of choice among options, which frequently lead to conflicts during evaluation.

Statements containing negatives are often ineffective and actually confusing due to the mental processing of the positive content prior to negation, "Don't talk" is processed as the command "Talk" followed by negation so confusion reigns. This matter, mentioned above, is extensively discussed in the Appendix (page 178) supported by further examples.

The bipolar state is usually generated by other people who provide a child, particularly, with contradictory messages. As **Agnes** grew up she enjoyed her Father's proclamations: "Such beautiful hair you have, what a wonderful child you are!" And she endured her Mother's: "It's time to trim that wretched hair of yours" and organized a trip to the hairdresser. Similar contradictory messages destroyed her sense of self by the time she was a teenager. She shaved her hair completely and suffered the antagonism of all but her age-peers.

The matter of polarity and contradiction comes up again in Chapter 3 where alienation matters are fully introduced. These concepts are effectively reifications; but are valuable concepts facilitating the discussion polar phenomena in terms of origins in mind in a similar way F. Perls used for his top-dog and under-dog.

Real and pseudo-emotion

Fear is fear of something – a scorpion or an imminent car crash. Fear of nothing in particular is pseudo-fear, an aspect of anxiety. We are familiar with the happy homosexual who is called gay because, in his socially oppressed state, puts on a bold pseudo-happy demeanour to protect himself. We also see politicians on the T.V. saying how angry they are about something or other, while showing no deviation from a usual, bland, unemotional state.

Paradoxes and double binds

Experimenters often produce paradoxical statements. **Angus** went to great lengths to explain how tired he felt when he went to bed at night yet he could not sleep. A paradoxical suggestion could be of the form; **T**: *So struggle to stay awake, read something, listen to the radio, do anything other than sleep.* A neat example of a paradox is given by Schoen[58] who noted that Krishnamurti encourages his reader to take no notice of teachers and therapists while he was himself pontificating as a guru.

Closely related to paradoxical situations are Bateson's double binds exemplified by the child who suffers contradictory messages from one or both parents. A father gently encouraged his son to speak up and make his needs known while mother continually shouted: "Shut up!" or words to that effect. The child sensed that speaking out and remaining silent will both be reproved if not actually punished. The child's decision process becomes frozen.

The adult in a state of confusion is similarly bound in indecision. We notice in therapy that this condition is pre-determined by the appearance of polar opposition factors and that concentration at this confusion impasse is followed by some kind of resolution, something completely different happens, comes into consciousness. This is often obviously a change of logical level as Bateson intimated. **Aldis** said: "I always wanted to join the Tory party and become a Cabinet Minister, if not Prime Minister, as my Mother encouraged me to do. But I found that I had great sympathy with working people and their state of poverty." In fact he did nothing about either cause. In group over many weeks resolution of this polar positions lead to an impasse and eventually to a conclusion and he began to work for ecological causes. He debated the issues in group over many weeks and did actually help the Green party in an election. He continues as an ecological aware activist.

In general the occurrence of paradoxes, double binds and polarity confusions are only disabling, particularly to a facilitator, if not sorted out for clarity, recognizing the polar factors, seeking via the impasse the dialectical solution that the experimenter can gain for self.

Two figures running simultaneously, or nearly so, can result in confusion – which figures shall have main attention and be acted on? – or the figures may have contradictory outcomes.

Now I am aware of … …

The impasse

The impasse is a hypothetical construct of value when discussing the alienated state people get into when they are unable to resolve personal difficulties, particularly the contradictions and dilemmas of polarity conflicts. The terminology is satisfactory since the person at an impasse becomes impassive. He is in a trancy state, lost to reality and gripped in confusing fantasy. Within the therapy group situation the myth may be something like: "They must dislike me for taking so much time." Once out of the impasse the welcome back is usually very warm. The jump out of the impasse occurs without really trying and has all the characteristics of the development of a healthy figure. The accompanying emotion may be anger, sadness, elation, Ah – Ha! and satori.

F. Perls and others make many references to the impasse with confusion due to lack of precise definition of the state. To clearly differentiate between apparently similar states of mind and body I use the following criteria. Looking at a person who is apparently quiescent, if he is –

- Happy and responds readily to communication: he is grounded.
- At peace with the world, probably with his eyes closed, respond gracefully and slowly to suggestions, then he is meditating.
- Eyes shut, absolutely quiescent, not available to suggestion and gaining attention, then he is asleep.
- Eyes shut, absolutely quiescent, but responds to commands, like arm levitation, then he is in a trance.
- Full of stubborn unexpressed energy after a cataclysmic concatenation of confusion, as when locked in a polarity situation. Then he is at his

own particular and personal impasse. He may respond dramatically, taking the opportunity to avoid confrontation with polarities.

What happens next?

The promise in the book title and the Introduction is that I will give attention to protogestalts and protofigures. The history in this chapter has been concerned with stuff that occurred fifty years before proto-anything. The next chapter gets nearer our main topic as the stages of gestalt formation are unravelled.

I feel stuck.
Just how is it for you, feeling stuck?
Difficult to say.
Do you accept that you have stuck yourself?

Notes and references

1   Ash, 1995. p. 85.
2   Köhler, 1947.
3   Wertheimer, 1945.
4   Ash, 1995, p. 73.
5   Rubin, 1914, translated by Ash, 1995, p. 179.
6   Köhler, 1920, in Ellis, 1938, p. 36.
7   Lewin, 1937, p. 283.
8   ibid, p. 290.
9   ibid, p. 250.
10   Wertheimer, 1945, p. 234.
11   Wertheimer in Ellis, 1938, p. 14.
12   ibid, p. 15.
13   ibid, 1938, p. 1,
14   Köhler in Ash, 1995, p. 184.
15   Koffka in Ellis, 1938, p. 393.
16   Wertheimer, 1945, pp. 75 → 84.
17   Polster and Polster, 1973, p. 10,
18   Maslow, 1967.
19   Perls often referred to moralistic regulation without clearly defining what he was on about. I infer that he was concerned with a relationship where the person makes judgements and gives or takes orders.
20   Perls, F. 1947, p. 189.
21   Wertheimer in Murray[59], p. 132.
22   Perls, F. 1948, p. 51.
23   Smuts, 1926, p. 230.
24   Perls, F. 1966.
25   Perls, F. 1969, p. 50.
26   Perls, F. 1975a, p. 93.
27   ibid, p 102.

28   Perls, F. 1973, p. 24.

29   Proto- seemed to be the most suitable prefix to indicate the precursor of a figure.

30   The differentiation of the two types of boundary does not occur in early gestalt essays resulting in some muddled discussion. [page 18]

31   Edwards, see chapter 7.

32   Latner, 1992, p. 24.

33   Lewin in Ellis, 1937, p. 291.

34   Lewin, 1948, p. 51.

35   Perls, F. 1947, p. 224.

36   Perls, F. 1969, p. 15.

37   Polster and Polster, 1973.

38   Swanson, 1980, pp. 71 → 85.

39   Gendlin, 1996.

40   Grove and Panzer, 1989.

41   Edwards, see Chapter 9.

42   Perls, L. 1994, p. 142.

43   Marcus, 1979, p. 42.

44   Naranjo, 1970, p. 33; 1980; 1993.

45   Simkin, 1976, p. 125; 1994.

46   Yontef, 1991.

47   Schoen, 1994.

48   Ziegarnik in Ellis, 1926, p. 314.

49   Wertheimer quoted by Ash, 1995, p. 295.

50   Perls, F. 1975, p. 15.

51   Soloman, 1995, pp. 190 → 193.

52   Perls, Heferline & Goodman, 1951, p. 51.

53   Perls, F. 1973, p. 30.

54   Perls, F. 1947, p. 145.

55   Perls, 1947, pp. 48 → 50.

56   ibid, p. 73.

57   ibid, p. 52.

58   Schoen, 1994, p. 202.

59   Murray, 1995.

Chart 1·2 revisited: see page 10.

People react differently on seeing this chart and report, in order of prominence,

1) four concentric narrow bands,

2) a cone sprouting from the page,

3) a conical hole penetrating the page, and

4) three concentric bands of twice the width of the narrow ones, or

5) a rather dizzying, rapid jumping between perception of several of these.

Whatever is noticed stands out as figure from the ground of the rest of the chart. This chart is constructed in similar manner to the many puzzle charts set out by Köhler[2].

—=(☆)=—

Chapter 2

Dynamics of grounded figure-stage processes

I first met dynamics
as an aspect of mechanics
and learned that everything moves and changes
unless stuck up against an immovable object
in which case each exerts"
the same though opposite
pressure on the other.

Resuming pursuit of the general theme of self-nurturing and its failures I find myself imagining that I am as facilitator sitting in front of an experimenter and putting questions to myself: 'What does this person need to do to adequately nurture self and solve his or her problems?' 'What clues are there to indicate how this person fails in nurturing self? Is it basic nutrition, food, etc., relationship problems, self-denigration, or inability to recognise that real life involves change or something else? Does he or she talk about anxiety, dissatisfaction, unhappiness and/or inadequacies in work and relationships?'

Having dealt with an exposition of the basic psychology of gestalt phenomena the present chapter will expand this approach to deal with recognizable parts of figures. A consequence of giving these points more attention will be to uncover clear stages of development and the general dynamics of the gestalt approach.

Before such exploration I want to refer again to Smuts[1] who drew attention to "stages in which holism expresses itself and creates wholes in the progressive phases of reality [which] may therefor be roughly and provisionally summarized as ..." and here follow 6 points. The first deals with the material world, the second with functioning as a living organism in self-maintenance, the third with central nervous control, the fourth with consciousness and personality and the fifth with social and ethical whole units. The sixth point is too complicated for me to understand.

So a figure is emerging, engagement and activity ensue. These involve changes as emphasized above, to self and environment. Before considering impediments to change I want to set out in detail the elements, as I see them, of the changes.

Presented in the Introduction was the concept that the starting ground of the gestalt figure comes from a neurological module pertaining to earlier memories of places, events and activities, the protofigure. A particular gestalt figure starts from its protofigure and finishes in the neural context as a new protofigure, with new memories augmenting those of the initial protofigure.

It is appropriate, when discussing the dynamics of gestalt figure development, to start with the protofigure.

Protofigures and memories

If we are able to know and to think about the network of neurons that constitute a protofigure, we would discover memories of past events and occasions that have similarities to a proposed new gestalt figure. The latter is thus primed with useful background information. Whilst this chapter is devoted to healthy gestalt figures it must be remarked here that the memory system may be very unkind if the background information is concerned with traumas and problems.

We have to leave this discussion at this point because the methods of finding out about protofigures belong in a later chapter. Meanwhile we can leave intrinsic affairs and explore in detail the extrinsic stages of gestalt figure development.

FIGURE STAGES

Where have I come from?
Where am I going?

At this renewed outset concerning gestalt analysis I emphasize again the unity, for the healthy person, of self and environment (field) in which all factors are intrinsic, activity, change, physiological, psychological, nutrition, respiration, social, cultural, historical and geological.

Lewin[2] exemplified the progress of learning as: "The practice curve, e.g., in typewriting, rises gradually to a plateau, proceeds there for a time, rises again, this time with something of a jump, and so on." In another, similar analysis[3] he considers the "Influence of time upon the effect of purpose ... and intention" noting that there are three "phases ... (1) conflict of motives, (2) choice or selection, (3) consummation," the latter being emphasized as action. This is essentially the process that Reich[4] represented and described for the sexual gratification process as a series of distinct stages culminating, if successful, in orgasm, depicted here as Chart 2·1. The Lewin and Reich circuits are paradigms for other circuits of activity –

- After a good meal comes a feeling of satiation and lack of concern for food.
- After beneficial exercise comes the feeling in the lungs of sharpness, acidity, exhilaration, of general pleasure and need for a rest.
- After a good shit comes the feeling of lessened tension in the anal and rectal area which disappears from awareness.
- After scratching an itch there is more than the absence of an itch, there is often a warm feeling of gratification.
- After a sneeze comes a feeling of clearness in the nasal passages.

F. Perls[5] says that he studied with Reich from whom he probably picked up an interest in cyclic activity. The starting place is always a rest position, a state of organismic equilibrium which Perls called the zero-point[6]. As awareness develops the first experience is sensory although this may be missed if the matter proceeds at high speed. Awareness of the need, for engagement with people, etc., follows, followed by mobilization of the necessary energy, by the action which results completion and satisfaction of the need. Withdrawal then results in return to the rest position, to the equilibrium state.

F. Perls described at first five[7], then four[8] stages of figure development (Table 2·1). Kepner[9] refers to information of this kind as being in the last chapters of Perls and colleagues book.[10] When counting I assume that Perls' first and last stages are effectively the same, ground. F. Perls[7] also described an "instinct cycle" (his quotation marks) as having "practical value [in] the enumeration of avoidances." He also called this circuit of events the "organism/world metabolism ... which consisted of six links" without, at that point, setting down what these were.

In 1948 F. Perls[8] was describing this process as "creating order from chaos; namely the figure/background formation." By 1951 he, with Hefferline and Goodman[10] redescribed these stages using terminology relating to contact, which

seemed to be his main concern at that time. They described these stages as "an account of the nature of creative adjustment and growth[11]" There is no sign that they consider that they are describing the life of a figure as part of a dynamic gestalt process. There are discrepancies between his various descriptions but at core the following stages summarise their ideas –

1) Forecontact: Where ground is the personal body and the figure is an appetite, urge, or "The given",[12] an environmental stimulus.

2) Contact: Where the ground is excitement and "the given" is accepted. There the figure is some object or set of possibilities. The self orientates and manipulates objective possibilities, is active and deliberate on body and the environment. There is affect. Identification or alienation occurs with choosing or rejection.

3) Final contact: Here there is no mention of ground. The figure is the "lively goal." Deliberation is relaxed. A "spontaneous unitary action of perception, motion and feeling occurs." The person is disinterested and concerned and fully engaged in the achievement of the figure. Awareness is very bright.

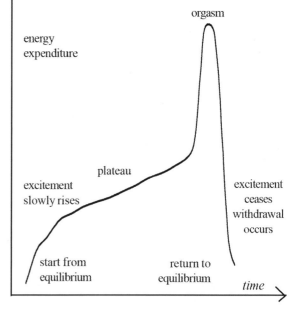

Chart 2·1 Reich's orgasm cycle[4].

Table 2.1 Development of the idea of stages in figure emergence and return to ground

Perls	Zinker	PGP — Organism	PGP — pathological	Frontispiece designation	This book stage
Organism at rest, balance, temporary, equilibrium	Equilibrium	Rest, fertile void, ground	Torpor, deadening void	Fertile ground	⟷
Disturbing factor, internal or external	Sensation	New organismic deficits or surplus	Dull incomplete deficits or surplus	Sensory presentment of need, participation	⟷
		Tension towards emergence of new deficit or surplus	Tension towards emergence of old deficit or surplus		
Figure, image or need emerges from background	Awareness	Awareness of dominant need real, unitary, flexible	Dull awareness, ambiguous 'if only' and 'must do'	Awareness of needs, etc. Thoughts about needs, etc.	◇
	Energy mobilized	Excitement, active arousal, moving, free breathing	Anxiety, restricted arousal, moving and breathing	Disturbed and diffuse in energy mobilize	◇
	Action	Contact firm and vibrant	Avoids contact	Engagement	⟷
	Contact	Experimentation, free ranging innovative	Items of old behaviours: no choices, manipulation	Action satisfying needs, etc.	⟷
		Action new, relevant assertive, complete	Action dull and irrelevant, confidence, etc.		
Gratification of need, Facilitation of development & growth		Satisfaction, discharge and assimilation	Dissatisfaction, depletion and self-interjection, 'indigestion', stability	Need is met and organism may be better	⟷
Does need function, gratification				Awareness of satisfaction of need and acknowledgement, stability	⟷
Foreground recedes	Withdrawal			Withdrawal	◆
Background	Equilibrium	Rest, fertile void, etc.	Torpor and further disturbance	Fertile ground	⟷
Return to rest, organismic balance and equilibrium					

Chart 2·2. Hall's representation of figure characteristics[14]

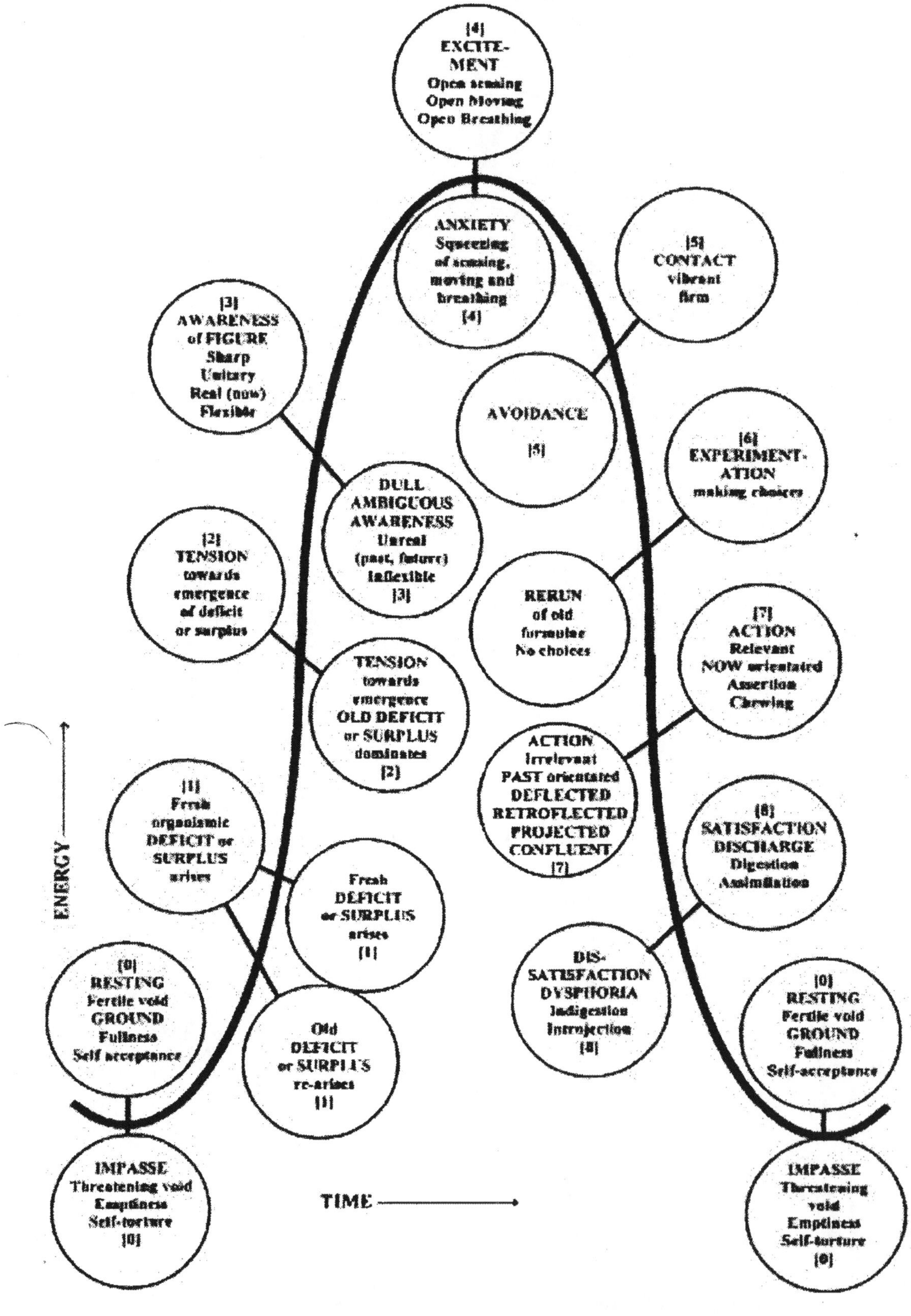

4) Post contact: The sense of self diminishes and a flowing organism/environment inter-action   occurs that is not figural.

There are indications in these descriptions that the authors considered that several figures were active simultaneously and associated with different grounds. While this is possible that at any given time several figures are active it seems to me that, by definition, a healthy ground – figure formation is one unit only.[13] Because of the lack of clarity in this 1951 essay the four stages are not summarised in Table 2·1.

Hall[14] and Zinker[15] became interested in stages of figure development and their ideas are summarised in Table 2·1, giving attention to a more dynamic way of viewing the emergent figure and utilized knowledge of the fine structure of the figure. While Hall recognized eight stages, and Zinker saw nine it is up to each person interested in the matter to decide how many are necessary for his or her purposes.

Diagramatization was an advance that clarified the concept of stages of figure development. Hall was interested in organismic flow and presented what appears to be one unit, approximately, of a sine-like wave with emphasis on the beginning and ending of the one figure in the ground. This is redrawn here as Chart 2·2. Hall clearly differentiated the stages exhibited organismically by the healthy person and pathologically by the person who interrupts the organismic processes.

Zinker was considering the creative processes of gestalt therapy. He pointed out, to take one of his examples, that interruption between action and contact (his model; see Table 2·1) could be interpreted as interruption of the flow of the events of figure formation which he drew as a sinuous wave. His continuous sequence of uninterrupted waves, representing what he called the awareness – excitement – contact circuit, is redrawn as Chart 2·3. Kepner[16] employed Zinker's form of diagram but substituted the concept of figure formation for that of awareness and referred to the whole as the cycle of experience and the cycle of self-regulation. The continuous, flowing, sinuosity of these depictions is misleading, as each gestalt figure is discrete with neither necessary dependence on pre-determined nor post-determined figures. As stated above, I prefer

to refer to a circuit of figure stages to emphasize the general occurrence of one cycle only.

Polster and Polster[17] described a "system for the contact episode [which] moves through eight stages" and these are essentially the same as Zinker's. They stress "This cycle may last for only a minute, it may play itself out in a session, a year or a life time. The eight stages … are guide lines and not to be taken as a cut and dried order."

In 1984 Tillett[18] produced an alternative representation of a circuit of events which did not involve grounding and introduced a role for equilibrium. This and other concepts and stages he discussed do not correspond with reality for me and I will not consider them further.

The representation, by Clarkson,[19] was circular and was drawn without emphasis on any particular stage. The inference from this chart was that each figure recycles continuously with no ending, and probably with no beginning. In the text Clarkson emphasizes the flowing nature of figures without attention to a time scale and without distinguishing the changing character of successive figures. The Clarkson cycle is redrawn as Chart 2·4. Most importantly, she made no reference to ground and grounding.

Parlett and Page[20] produced a nine-stage model splitting the contact point into "engaging fully" and "final contact." These are all essentially modifications of Zinkers cycle with action preceding contact. Melnick and Nevis[21] employed the same schema in the context of considering DSM-III diagnoses and Carlock, Glaus and Shaw[22] similarly in work with alcoholics.

Rose[23] has emphasised the importance of recognizing the interrelationship of cycles, short term ones, within the compass of long term evolution. She emphasizes that it may be necessary to encourage completion of short term matters in the context of long enduring processes. Wheeler[24] reproduced Zinker's cycle. Mitchell,[25] had cycling in mind as he worked in therapy with a three phase model – an opening which "created space for dialogue, a middle phase where "mobilizing energy, exploring choices and beginning to engage" occurred and a final phase of "action, satisfaction and withdrawal.

Chart 2·3. *Zinker's depiction of figure characteristics.*

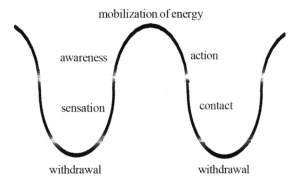

mobilization of energy

awareness   action

sensation   contact

withdrawal   withdrawal

Chart 2·4. *Clarkson's depiction of figure characteristics.*

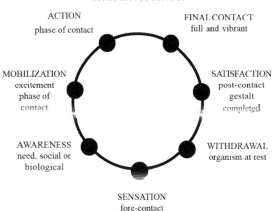

ACTION
phase of contact

FINAL CONTACT
full and vibrant

MOBILIZATION
excitement
phase of
contact

SATISFACTION
post-contact
gestalt
completed

AWARENESS
need, social or
biological

WITHDRAWAL
organism at rest

SENSATION
fore-contact

Cycles, recycling and circuits

So growth and creative adjustment to self and environment, to expand the above quotation[27], occur with well defined stages for figures. A figure sprouts from ground, does its job, and regresses to ground: recall as a memory is then possible as is influence on similar figures in the future. This can be considered to be one cycle or circuit, as suggested above and as Hall[14] says.

Zinker[15], however, considers that cycling goes on and on – in his representation his waves have no beginning and no end. Clarkson[28] in depicting figure-life as a circle also has no regard for time limits. There is a polarity issue here. One cycle or many. Examples make the matter clear. At one o'clock the church bell strikes once but for a funeral the sexton tolls the bell many times. Both propositions are valid according to context. However, for simplicity, I will continue this chapter with my eyes on the stages of one grounded, complete healthy gestalt figure.

In everyday, wakeful life figures follow one another all the time even with a degree of over lapping. While meeting a friend is one figure our Queen meets a line of dignitaries and speaks and behaves differently with every one of them – I hope!

The single figure of my interest, then, in the following pages, mostly consists of one circuit of stages. So it is convenient to stop considering them as a cycles and refer to it circuits of stages.

Depiction and representation

I was convinced by the foregoing authors that the generalized life of a gestalt figure had complete, discrete stages which provided an accurate enough map of events during figure formation and dissolution. All figures are essentially of the same general pattern with marked differences only in the character of the various stages. I don't totally like representation of the stages in such unique, discrete, compact ways because I am aware that although sensations occur at a certain stage, they continue, develop and change character as engagement and action supervene as the dominant characteristics. While it is obvious that engagement must occur before the need satisfying action can occur, it is also obvious that engagement continues during the action stage and that both end when satisfaction occurs.

Chart 2·5. A student's problem.

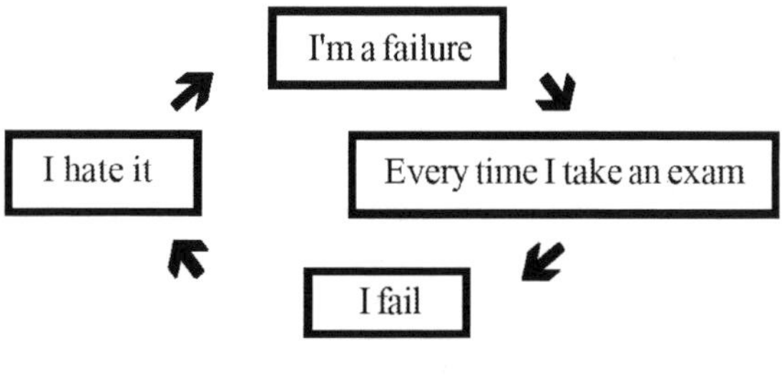

However, there is enough differentiation at each stage to enable use of the stages as an analytical tool.

What people say often indicates a staged process. **Bert** says: "I am anxious about becoming more anxious. Then I realise that I am more anxious, and so it goes on." **Alice**, somewhat cynically joins in and says: " I enjoy anticipating joy and become more joyful." **F** says: *"What is the opposite of anxiety for you,* **Bert***?"* No reply. *"Does anything you do increase your sense of anxiety?"* Silence. *"Is there a particular aspect of your anxiety that interests you?"* **Bert** says: "I am always anxious. Like now I feel tetchy at your questions. Yes, I'm down right angry that you don't leave me alone. I need time to really get to know how I feel when I'm anxious. People are always interrupting me. I'm on edge all the time, ready for the next blundering fool." *'What do you say when a fool blunders around you,* **Bert***.'* Nothing much, I can't be bothered." *"So you stay anxious."* **Bert** is stuck, fussing at the beginning of his figure formation.

Bey says: There always seems to be something around that makes me feel afraid. If I'm home I suddenly remember that I'm afraid of going out where I will become afraid so I feel uneasy at home, too. *'What is the opposite of fear for you.'* **Bey**, after a few minutes: "Courage, I would like to think that I will become courageous." *'Can you remember a time when you were courageous, etc.'* This cycling interest, anxiety working on anxiety, fear on fear, is only really troublesome when escalating occurs; depression compounds depression. In absence of escalation there is a good chance that the person will find their own way out of the circuit and not present for therapy.

Boris says: "I want to get qualified so that I can take a senior position in my job but I keep on failing the examinations. Every time I take the exam and fail I feel a worse failure, I hate it. I know I'm a failure and I might as well give up." Since he has taken many exams many times he is operating in the circuit depicted in Chart 2·5. Although he knows what is going on he has not recognised that the events form a repeating cycle. This lack of awareness results from the fact that the individual incidents are spread over a longish period of time. Inclusion of a time scale could pull the cycle out to form a spiral. Representation as a spiral in three dimensions, rather than a reiterant cycle, could also be set out in catherine wheel style, with the time scale exploding outwards. If energy is similarly represented reversal could result in depiction as an imploding, involuting spiral. While these are interesting representations (maps) of ideas and observations they will not be further explored in this book.

My own summaries of observations and ideas are based on use of Ockham's rule[29] that may be stated as *"among answers to a question the simplest is the one most likely to be true."* I prefer an alternative shape for my chart to emphasize the base character of the ground and, with Lewin and Wertheimer, the apotheotic, consummatory nature of *activity* as the pinnacle where energy expression is generally highest and engagement most firm. Again, L. Perls[30] says "… coping with the **other**, the different, the new, the strange … is not a state we are in or out of … but an *activity*.

(L. Perls' emphases). My complete and fully elabo rated chart is presented as the Frontispiece.[31] This chart forms the main algorithm of this book and will be modified to illustrate further arguments; the stages will be generally depicted in the form <X>.

Thus my perception, as will be elaborated later, is based on Hall's depiction of engagement preceding action and using a time scale. Zinker's highly influential model, where the main action precedes contact, seems to me to be passive rather than assertive. As for the nomenclature, it largely does not matter. Experience is figural so the circuit of figure development is also a process of experience, whether on a micro- or macro-scale with respect to time. However, my representation not only recognizes the discrete stages within the circuit, it also emphasizes the uniqueness, the unitary form of each figure on the basis that every healthy figure of every kind flows through the same format and does not necessarily repeat, as elaborated above. This seems certainly so to me provided that I can recognize that the speed of flow in figure development may make it impossible for me to be aware of certain stages as being present at all. The relationship of Stages <1> and <2> is the most obvious example; I am knowing and verbalizing <2> about a sensation <1> almost before it has occurred. Stage <7> is often missed because busy people do not stop to recognize it. They get on with the next figure. In tennis my arm and racket are reaching (stage <4>) for the ball before I hit it (<5>).

The energy, Lewin[32] and Goldstein's tension, hangs around if there is no completion. "Activity units often display striking initial and end characteristics (e.g., a sigh of relief, an emphatic gesture …) which serve to mark off the boundaries of a temporally unified field." For Lewin, says Ash,[33] "The unity of perception and action was one of the cornerstones of gestalt theory. Thus it was only natural to transfer gestalt principles to the study of action and emotion." These dialectical unities are essential in any theory of figure development as will be discussed later.

Stages in the development of a figure

Recognition of the figure is central in the theory of academic gestalt psychology. And, as Wertheimer stressed (see Ellis[34]) the dominance of a major, powerful figure may make it difficult to be aware of sub-figures. However, presuming, for now, that a figure develops without impediment in the healthy state, we can proceed to examine the detail and dynamics of the fine structure, referring to Table 2·1 and to the Frontispiece for a summary of the stages. The diagrams of Reich, Chart 2·1, Hall, Chart 2·2 and Zinker, Chart 2·3 represent the passage of time in an approximately linear manner. Those of Parlett and Page,[20] Melnick and Nevis[35] and Clarkson, Chart 2·4 do not. It seems to me that it is essential to recognize and represent duration, grounding, energy status and passage of time in any depiction. Fuller descriptions of each stage[36] are now set out together with the reasons for distinguishing each stage from its neighbours. Some indications of the relevant affective associations are given as are some allusions to rudimentary aspects of events of the corresponding unhealthy state, though most of these are set out in Chapter 3, because, I emphasize again, I am mainly concerned in this Chapter with the healthy state.

The sections on the following pages are accompanied by stories which develop stage by stage and are designated ♦, ♥, ♣ and ♠.

<0> The fertile ground

The featurelessness of the organism at rest, grounded, in a state of centredness, integration, poise, balance and equilibrium, non-sensing, inactive, relaxed, restful, and a meditation state was described by Hall[14] as the fertile void. F. Perls used this term but did not, as far as I can ascertain, equate it with the ground state. Perls frequently used the term zero point. It is clear that Perls used the term void in association with pathology so it is not appropriate to use it here. I prefer to celebrate the potential for growth of the healthy state by attaching the quality of fertility, to define the fertile ground. One of the aspects of fertility is memory, the source of information gathered in the past. Attributes of memory that are of vital importance are the emotions and the propensity to develop further.

In my poised state I am ready for any eventuality, whether I initiate it or not. While the dynamic gestalt is as yet nonexistent there may be the sense of disengagement and independence that Wertheimer wrote about (Table 1·2). This is where Lewin's "psychical energy (as a result of some tension or need) – i.e., psychical tension systems" occurs.[37] Also[38] "systemic equilibrium does not mean an absence of tension in the system for equilibrium may also be attained in a state of tension." A standing person has just enough tension in the back and front musculature to maintain poise and stillness. This is also Goldstein's "mean point" to which the excited organism always strives to return (Ash[39]).

As suggested on page 33, the memories of past events constitute protofigures, the seeds of the gestalts that will blossom in the environment.

<1> Sensory presentiments of need for self-nurture, etc.

Knowledge of the sensory impression, a feeling of tension, an object seen, thing heard, something smelt or tasted, something touched or internal impression such as hunger, thirst or an itch, clearly differentiates such phenomena from the ground of no-sensation. At this stage, there is no cognitive, verbal description of what is going on and yet this stage is the essential prerequisite of being human, being a vital self. However this is a fleeting impression in many circumstances. An unfocussed apprehension, a vague lack of sensation and affect may also occur with ideas on thinking about it, about dissociation and the trance state.

This is the point of Wertheimer's (page 16) emergence, where interest "springs forth" from the "not as yet formed" (Table 2·2). This is LeDoux's "preattentive, low level process" stage. This is where the objective of self-nurturing is realised.

F. Perls suggested that the earliest aspect of the emergent figure that could be recognized was a pre-cognition, a "disturbing factor." This is a sensory impression as Zinker[15] acknowledged, that then becomes recognized in awareness as a new stage. In terms of water-lack, the sensation of thirst is recognized before the internal or externalized verbalization: "I am thirsty."

Hall[14] described this situation with inadequate specificity. He recognized what he called a "water-deficit" which is really recognition of water-need, again usually called thirst. He then attached an extra abstract concept, "tension towards emergence." This brought in an unnecessary factor, a sort of "life force." It is enough to recognize what is well known, that sensory impressions precede *awareness* of the sensory impressions, when such differentiation is possible (as said above, the possibility is not available in the high speed situation; the throb of pain and the idea "tooth ache" may superimpose). However, a degree of desensitization is essential in some ircumstances. Thus if I am reading a book I would rather not be aware of a neighbour's TV or an itch on my foot, or any other intrusion. Sensitivity is essential as a figure develops – it may with benefit vanish as time goes on.

It is the most vital need that gets the attention and it is sometimes clear that there has been a hierarchy and one becomes dominant.[40]

The energy expression at this stage is minimal. The musculature is not involved, only electronic pulses in the nervous system connect with thoughts and feelings.

♦ I have a fleeting, fantasy impression of my garden where I could walk.

<2> Basic awareness of need for self-nurture, etc.

Nothing happens without the presentiment
of the possibility of it happening.

When tuned to sensory impressions I experience visual, auditory, olfactory, gustatory, affective and tactile impressions and I clarify knowledge of events for myself by verbalizing, talking to myself or someone else about what is going on. I also have, somewhat hallucinatory impressions as an idea or fantasy, as, for example, of food if I feel hungry. Such awareness is cognitive and differs qualitatively from sensory impressions: this is justification for knowing these as clearly different stages in the development of the figure.

This, for Wertheimer was the emergence of meaning and

intention. For him there was no chance about this – everything that happens in a figure facilitates development of the figure (page 16). For F. Perls this cognitive stage, awareness of need, was the first stage of the recognition of the existence of a real or imagined figure. Zinker[15] and Hall[14] knew this as the stage at which the dominant need, the need of the moment, forms the figure. For Hall this is always a real entity whereas, in contrast, Perls also admits fantasy imagery. I go along with Perls since it is well established that guided fantasy approximates to reality in therapeutic efficiency.

Hall adds some valuable qualifiers to his description of the need awareness process. He adds "unitary"; his qualifier is most apposite as he stresses that, in health, we are concerned with one clear figure at a time. He also adds "flexible" and thus stresses fluidity as a characteristic of the need for satisfaction.

While the description here is of awareness of need and appetite, it is necessary to add to physiological need, the more generalized concept of want and the less energized states of curiosity and interest. Then when fully aware I know that these tensions can be considered as the drivers of figures. Also, the vague lack of sensation referred to at the end of section <1> above leads to recognition that something is lacking and desired and that there is a need to do something about it. "What do I feel and know about this, whatever it is?"

Simple awareness of need, etc., usually takes little time as self-stimulation develops. If I breath into feelings to intensify them I know that I am alive. However, supposing that person **X** recognizes a need, there is no guarantee that he will do anything about it.

While the impetus may appear to be need, etc., as set out above, behind the scene are beliefs, memories, hopes, intuitions and the general potential of people to pretend and experiment.

While there may be much internal and external talk about needs it is the need with most vitality that actually gains attention. Here cognitive processes are of value in ascertaining priorities.

Stage <2> provides a choice point of awareness. A procedure for sensitizing awareness is described on page 164.
♣ *I carry a bow and arrows and I am stalking a deer and so is a wolf – shall I go on or be safe and back off?*
♦ *I want to go to the garden.*

<3> Excitement and the mobilization of energy

Nothing happens without a modicum
of enthusiasm.

Here, enthusiasm, excitement and energy augment awareness, affect and ideas. Concepts about activity form a clearly differentiated phase differing from the conceptual stance of simple awareness and options for action may begin to be considered. The energy involved is one or two orders of magnitude greater than at stage <2> as decision turns towards practicality. (A resolve does not carry as much energy and decision and motivation are even weaker). Then occurs a very pleasant state of being in which anticipatory excitement may continue for some time as with lads and lasses getting ready for a dance. There is much pleasure in prospect and this is also similar to the manner in which some people anticipate pain.

Wertheimer (Table 1·2) described the emergence of expression, targeted on satisfaction of a goal. He also emphasized the link of affective processes and awareness. This is where Lewin's[41] activating energy occurs.

Zinker[15] considers the appearance of energy at this stage while Hall[14] gives main attention to excitement. Both are energy phenomena and in general, in so far as the process is slow enough to enable recognition, excitement precedes all specific affective and other energy manifestations, crying, laughter, anger, fear, running, etc[42]. Some degree of change of breathing and excitement is usually the first indicator of any affective display.

Energy and affect are fruitless in themselves. A next stage is necessary. The energy around affect and excitement has lead some writers to recognize this as the action stage. It is clearly not so because there is as yet no engagement with what is required to satisfy the need. And the satisfaction of the need is not yet happening.

♦ *I walk towards the garden.*

<4> *Engagement[42] and mobilization of resources*

Engagement is the necessary preliminary
if something is to happen:
"Get in gear, Man!"

Here required resources are mobilized ready for the activity utilizing the resources to satisfy the needs, etc. F. Perls and colleagues[43] say: "We use the word contact … as underlying both sensory awareness and motoric behaviour." Engagement is a word of clearer meaning than contact, is instantaneous in effect and is only meaningful if followed by action, work, assimilation, growth, etc., that are temporal activities.

Engagement is the first actual activity of the subsequent sequence of effectively fulfilling needs, wants, etc. and occurs on the self-environment interface – out there the action can occur. Meanwhile, back inside myself, I can investigate what it feels like to be on the border. Engagement is not simple excitement because the figure has now got to the point of interaction towards achieving satisfaction of the need and yet is not doing so. In a kiss two pairs of lips contact before the full activity of kissing develops. When people meet "Hello!" is the usual contact word though there are many others including "Hi! Fancy meeting you!"

When I am in really effective engagement, contact is assertive and firm (and not timid or forceful).

"Who is this person talking to, me, self or the wall?" Contact as a word indicates a static state: "I am in contact."

"Gestalt is the appreciation of differences" said F. Perls[44] about the boundary between self and other. "I like …." "I don't like …." "I go on to be active or stop, inactive and isolated." Real engagement and association with people are cooperative, there is no compliance, submissiveness or deadness. The only engagement worth considering leads

instantly to activity and is often instantaneous after the "psyching up" of stage <3> and just as the action is about to get going. Engagement involves a small degree of preliminary activity as when I pick up my concertina ready to play and I am not yet playing. An artist takes up a brush loaded with paint and is yet to apply it to the canvas.

Wertheimer, Lewin and Goldstein do not seem to have attached any importance to engagement as a phase of gestalt development. I have found it important, when clarifying my thoughts, to separate contact and action as separate stages. Both Zinker[15] and Hall[14] add contact to F. Perls schemata and Hall emphasises that effective, real contact is firm and vibrant. His description of energy charge is also useful.

From and Miller[45] say that the: "contact boundary is where growth occurs [and] gestalt therapy is concerned solely with activity at the contact boundary where what goes on can be observed." These are erroneous statements because growth is dependant not on engagement but on activity. Perls and colleagues would agree with me on this aspect because they point out that self-discovery is the "observation of self as action."[46]

In now crossing the boundary all the factors concerned with adventure and risk operate – these were detailed in Chart 2·1.

♦ *My hand grasps the handle of the door to the garden.*

♥ *The thirsty person can have a glass of water against his lips and yet not drink. Or a mouthful of water may be swallowed and promptly vomited; no engagement was then made with the internal water absorptive processes and the need fulfilling action of absorption did not occur.*

♠ *The runner is on the track, foot on the block, waiting for the starter's gun.*

<5> *Action, Association and consummation*

The richness of life
is measured by its activities

Aristotle

Action is life. "I eat, drink and am merry." Here fantasized engagement becomes real, affect is fully expressed, and thoughts empowered. The trance of engagement becomes touch; skin to skin, word by word. Action breaks isolation. Here, we are, out in the world! "I like you." "I light a fire!" "Here I am nourished."

Action is a move on from fussing around with preliminaries, is the point of consummation of need, want, etc., is the apogee, the apotheosis, is an active process and needs a dynamic word – activity in fact. It is what is happening. It can even be orgasmic. (Chart 2·1). Lewin[47] says: "this tension system [stage <4>] thereupon assumes control over motor behaviour" which then "leads to satisfaction and resolution of tension." He further emphasizes[48] that this is his last phase, that of consummation. My Mother's old Kentish saying was: "Actions speak louder than words."

At this stage the activity, as Hall[14] emphasises, is: "Effective, economical action is relevant, free ranging, innovative and assertive. Activity deals with the definitive need and is the stage of involvement with satisfaction. Activity connects past interest and future satisfaction."

When people meet, their mutual, cooperative business follows the contact word "Hello!" Activity is directly related to the initial stages <1> and <2>. "What shall we do now?" However, beware of gestures, the imitations of activity so well described by Genet. When it comes to a particular activity, is there a rule about how it is to be carried out? Must it always be done in the same way? If I am prejudiced I have no room to manoeuvre.

Being totally engrossed in activity was contrasted to contact by Simkin.[49] He suggested that "You have lost your contact function" and "lost your contact boundary," an erroneous statement since activity can only occur while engagement continues.

Kepner,[50] who considers that the need satisfying action precedes contact, refers to his equivalent of this stage as, the after-action. Not so. When the job is done it is time to tidy up, to log-off and get ready for the next figure.

♦ *I turn the handle, open the door and go into the garden.*
♥ *The thirsty person drinks down the water.*
♠ *The gun explodes and the race is on.*

A man becomes a man
by dancing with his people.

Janacek

<6> *Engagement is broken off as the need for nurture is met*

The happening has happened
farewell!

Once the need, want, etc., is met, engagement can be broken off as excitement and energy expenditure die down. Wertheimer (Table 1·1) emphasizes his interest in achievement of the whole, a totality, as his ending. This initial target had been achieved.

For F. Perls this is the conclusion and "the decline of the disturbance," as he describes it, of figure formation, of engagement and of action. When action has ceased and there is no hanging about. A new feeling and ideas may begin to arise towards the next need satisfying action figure.

♦ *I am enjoying the garden.*
♥ *I no longer feel thirsty.*
♠ *The race is over.*

<7> *Satisfaction and validation*

As I finish what I am doing
I become well pleased with myself.

Once the action is over a change of awareness occurs: the need is satisfied. At this stage excitement and energy expenditure have practically ceased and only the cognitive state of awareness pertains. Acknowledgement of satisfaction, validity and pleasure form a complementary state that contrasts with the earlier awareness of the need. The self has indeed become nurtured.

As Lewin[51] says: "Satiation involves not only a change in locus of these forces [e.g., at stage <5>], but also a decided change in the psychical tension which had underlain the goal-seeking behaviour." For Wertheimer both emotional and intellectual satisfaction have been achieved.

With the feeling of satisfaction, as Hall[14] observes, comes a feeling of discharge. F. Perls[7] commented on the decreased tension and the feelings of discovery, of gratification and compliance. The "Ah-ha", "Wow!" and satori experience may occur at this stage.

In the run of everyday life it is rare that one pauses and is aware that success and gratification have occurred – "That's better!," "Good going!" – and that these have life affirming validity that are accompanied by good feelings. Such a summing up is also a valuable part of cognitive experience because every successful run of the organismic circuit returns, as Hall[14] emphasized, to augment the fertile ground. From this enriched ground new fecund figures arise.

One of the dynamic expectations, as a successful figure ends in satisfaction, is that a new exciting, successful figure is about to begin; a triumph of awareness. Some of the words associated with this position are good bye, 'bye, adieu, so long, ciao and farewell.

It is possible to consider satisfaction and validation as two separate processes. However they are intimately related cognitive, idea and affective processes and I now regarded them as two aspects of the one stage.

If satisfaction does not occur desire continues without end and is, say the Buddhists, the origin of suffering. Feelings of satisfaction are, in effect, rewards for beneficial behaviours that occur without prompting. It is a good idea to reinforce them voluntarily.

♦   *I am in the garden where I want to be and I am content.*
♥ *Whatever happens next is a new figure.*
♠ *It was a good race even if I did not win.*

<8> *Withdrawal*

Nothing succeeds
like success.

Old Kentish saying.

The purpose of the emergence of the need figure is now fulfilled, self-nurturing has occurred and closure with withdrawal to ground occurs: this is a natural retrogression and recession (Table 1·2). While the healthy person enjoys a sense of validation of the experience and a glow of satisfaction, there may also be a sense of tiredness and a need to relax, even to sleep.

Withdrawal is differentiated from stage <7> because, under some circumstances, minor activity is necessary to round off the action, such as cleaning and putting tools away after creative craft and art work. Though it must be admitted that such tidying up can also be considered as a new figure. Another reason for differentiation is that some people get stuck at stage <7> and do not move to ground, as will be considered when discussing pathological processes. When I complete my business I let go ready to get on with something completely different. A neat final question is: "Have I done what I set out to do?" The statement: "Now that's that" with smiles leads on to something else next. A new figure may rocket off promptly and I may be so pleased with myself that my next figure has me going to the drinks cupboard for a tot of whisky in celebration. My paternal Grandmother always said: "There" or "There now" with a falling intonation and a very satisfied, happy face. I knew she was pleased with what we had done.

Then begins a readiness for a new figure and the sensations of that figure may begin to arise for a new stage <1>. And I can say: "This is me! I have done well, done a good job."

A good exercise for practising complete withdrawal is to play the Patience (Solitaire) card game. A new game will be confused if there is a residue in mind of the numbers of the previous game.

Active and out of awareness assimilation occurs with "Didn't I do well? I look forward to doing this again!"

<0'> Ground

Grounding is reached after fulfilment of all the steps of the circuit and the personal gains of the beneficial experience. The ground is the memory system and may be significantly altered in a material sense, as when a work of art or craft is created. For better or for worse, the ground is not as it was when the figure began to emerge because, as set out above with reference to Hall[14], a new beneficial experience has become added to all previous experiences. Under beneficial circumstances the grounded person has withdrawn to a centred, fertile, poised and integrated state, ready for the next venture. This total figure experience is remembered and becomes a factor in all future figure development. When grounded again I can be back in a state of equilibrium, I am inactive, resting, centred and relaxed, poised and ready for any eventuality.

Assimilation of the benefits of an experience continues out of awareness and the nub constitutes a new protofigure that lurks in the ground waiting to be reactivated when an internal or external stimulus engages with it.

Thus the organism returns to systemic equilibrium (Lewin in section <0> above) and to its mean point (Goldstein). The kiltered, healthy, need satisfying, engagement and action circuit provides both an account of figure formation, a powerful analytical tool and a protofigure, as I propose to exemplify in what will be accomplished in the following chapters.

One good outcome
leads to another.

Hindsight

An alternative way of looking at the figure stages is to examine them retrospectively. F. Perls[52] did this when he described the pre-cognitive state, now designated stage <1>. This way of seeing the stages may be useful under some circumstances, such as when teaching. Comprehensively –

<1> precognitive
<2> cognitive, pre-energized
<3>   energized pre-engagement
<4>   engagement pre-active
<5>   pre-terminal active
<6>   pre-satisfaction completion
<7> pre-withdrawal satisfaction
<8> withdrawal

This is not, for me, a neat way of labelling the figure stages. There is too much likelihood of muddle.

Qualifications

I want to emphasize aspects of my introduction to this chapter that were set out on page 34. The naïve picture of development of a figure, set out above, places all emphasis on outstanding events at unique stages, for example, that pleasure occurs with satisfaction <7>. This is obviously not only so. A healthy person will have gentle pleasure during figure development in anticipation of the climax at termination. A carpenter making a chair hums or sings his pleasure while working and ultimately chuckles as he sits on the completed object.

Other affect may be similarly generally evident although the main manifestations will be at stages <3> and <7>. Remembering Reich's observations, Chart 2·1, the pleasures of foreplay become magnified and overwhelming at consummatory orgasm. And when a trauma errupts deleterious affect dominates at stage <1>.

I have emphasized above that I find it necessary to consider that engagement occurs before action. Other people see this the other way round. I am an action man, rather than a sedentary man, and this no doubt pre-destined my choice. Action is consummation and surely cannot occur without an initiating engagement!

Phases of the figure processes

Here I begin to interpret Lewin on the "practice curve" as discussed on page 34. He clearly knows that tension <1> was associated "followed" would is too mechanical a word – with choice, a knowledge state <2>. His valence was directive towards action <5> with consummation which was about in the environment. Satiation accompanied by loss of tension then occurred <7> I can go with Lewin and feel that withdrawal to the fertile ground is achievement of an equilibrium state.

The depiction of the Frontispiece separates events beyond the personal environment interface from those within. Stages <4> to <6> from stages <1> to <3> and <7> and <8>, emphasizing the engagement and action phase.

An alternative approach can single out another biphasic view in which stages <1> to <5> can be considered to be of creativity and positivity while stages <6> to <8> are of dissolution and negativity, not a destruction phase, as F. Perls exaggeratingly called it. Hence both positive and negative aspects are essential dialectical characteristics of the figure – it must be negated to release attention and energy for development of a new figure.

Self and the figure-stage processes

Accepting the generalized stages of figure development as set out above, what impingement then occurs on the theory of self and mind? Further discussion will be clearer if set out stage by stage –

<1>   Self needs a piss – this is a deep seated sensation in the area of the urinary bladder, and is known by mind.

<2>   Self realizes that he needs to piss. The idea reverberates in his mind as he works out where the nearest toilet is situated.

<3>   Self mobilizes energy and may feel trepidous – will I get there in time? – and gets moving in the direction of the toilet.

<4>   Self reaches the toilet and unzips (it has been simpler to consider the trousered predicament) – mind follows progress.

<5>   Self pisses.

<6>   Self has lost the tension, anxiety and sensations of a full bladder and is no longer worried about reaching the toilet in time. Mind knows all about this and remembers where this convenient toilet is situated for future use.

<7>   Self and mind know that he is satisfied and may well feel pleased with himself.

<8>   Self begins to wonder what happens next.

<0'>   Self is ready for whatever happens next.

This example illustrates that self is responsible, is the instigator and executor of figure development. Mind knows what is going on and provides a commentary, information – he knows where toilets are located.

This is operation from a firm centre: any other condition (as discussed in Chapter 4) would behave differently – he might even foul his trousers as a means of gaining attention. However, there is still a measure of adaptation since pissing on the floor does not occur. Life is more comfortable that way. A lesson was learned at an early stage of life. A well adjusted male, in a public toilet, is also gregariously, conspicuously standing alongside pissing neighbours.

Self is thus the active person who goes out into his environment to satisfy his or her needs and finds nurture. An experiment to sensitize awareness of the stages is set out on page 178.

Figure life

If a figure repeats, even with minor modifications, it can be considered to be recursive and truly cyclic, as emphasized above. Most figures are not repetitive, they are one off and it is not helpful to consider them as cycles. It is more appropriate to refer to them as figure-stage circuits, as suggested above. An example of the former is the action of a man sawing wood with many strokes of the saw and of the latter is shutting a door in one go.

The time span of the run of a particular figure may vary from very short, as with an itch (Chart 2·6), to very long as with a course of study stretched over years, illustrated in Chart 2·7 where inner recycling can be considered to occur because interruptions by the general processes of life occur including eating, sleeping, chatting, holidays, etc. Resumption of progress of the figure occurs, at the point where it was left. And as stressed above, longer term figures may contain sub-figures and sub-sub-figures. All this is part of complete awareness of what is going on. Thus the construction of a house is a large figure. Sub-figures include laying the foundations, building walls, placing the roof, glazing, painting, etc., sub-sub-figures include making bricks, preparing wood, etc. For a builder the house is a sub-unit in a housing estate. Gestalt analysis of these logical levels can only be itself a complete figure if these factors are taken into account. Gestalt analysis is aided by use of charts if they clearly illustrate interruption processes and their outcome.

Figure life correlates with time. Action, stage <5>, can only occur here in this particular now as present tense activity: "I am going." When the action is finished, at stage <7>, one can look back at the finished action and past tense applies: "I went." The future looks bright: "I will go!" "I intend to go," "I choose to go," defining conditions for an up and coming figure.

Here, incidentally, is the rationale for F. Perls' "Be here, now" dictum. Therapy is itself figural and can, like any action figure, only occur in the present time continuum. So it is rational to bring the past and the future into the present using fantasy artifice, exemplified by present tense verbs. An obsession with "here and now" is, however, hopeless technique for every day conversation and life in general.

Sentences are themselves figural. The important parts

are verbs for action and noun subjects or objects. Various extra bits qualify and elaborate what is going on and in all, anything larger than the simplest sentences show personal intention, awareness, engagement with an execution of action followed by completion and, hopefully, satisfaction and expression of happiness.

In the model I have presented aspects of need, or something associated with need[53], feature at the origin of every dynamic gestalt. While I have conceived of actual needs being physiological and wants as being more or less intellectual, appetite, impulses and curiosity attach to both need and want to some degree.

A further inspection of the nature of need was set out by Maslow and described as a hierarchy of needs as indic-ated in the Box on page 14. Maslow's proposition has much in common with the proposals of Berne who, however, did not rank the needs. There are aspects in common with the procedures I was taught as a Student Counsellor (Mallinson) – make

sure that he or she is properly fed and has somewhere to sleep as if at home, that he or she has companions as is easy in a hostel but sometimes missing in digs. Then attention can be given to presenting features such as fear of failure on the study course.

I ask myself: where in the figure-stage scheme of things do I think? Thinking is the verbal expression of awareness, which occurs first at stage <2> and then stage <3>. Thoughts at stage <4> onwards are like: "I'm busy, getting on with the action and do what I want to do. This is fun, doing this." Then at stage <6>: "So I have done a nice job – what comes next?" This is all stream of consciousness thinking which occurs all the time. It may carry very valuable information: "Be careful as I pick up that pan, it is hot!"

Thinking can accompany any stage of the circuit except <0>, <1> and <0'> and is most prominent at stages <4> as engagement occurs and <5> during activity. All this is independent of the occurrence of affect which begins at stage <2> and may climax at <3>. The nature of the affect may change dramati-

Chart 2·6. Dealing with an itch.

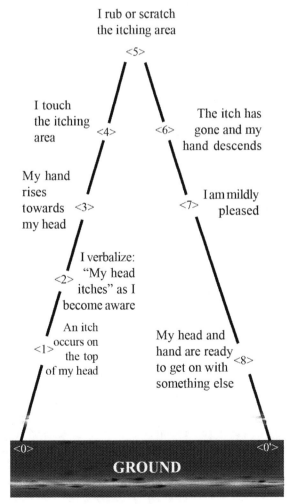

The total transit time is probably 2 to 5 seconds.

Chart 2·7. Becoming a graduate.

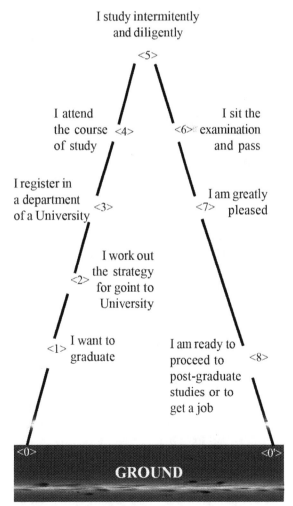

The total transit time is probably 3 to 4 years.

cally with stages <4> and <5>. Excitement at the prospect of meeting a lover may change to deeper feeling of love on meeting and then frenzy when in bed. After stage <6> less energetic feelings ensue as satisfaction sets in. At stages <3> and <4> I may become concerned about options – "Shall I do it this way or that?"

Thoughts may change as rapidly as affect. I know that I am happiest when my feelings and thoughts are in concert. As I think about thinking I become aware that many of my thoughts pop up in serendipitous fashion. They come from memories of my past experiences and parental messages (injunctions, Dawkins' psychoviruses, rules of behaviour). Many are beneficial and many others may need to be rejected out of hand. So I select among thoughts in a way that I cannot select from among feelings. I censure. I remember and still leave the memory to be remembered again another time though in remembering I may modify the memory.

On thinking about my itch (Chart 2·6) I realise that if I am chatting excitedly about something else the whole process can run with minimal awareness and minimal thought. That would not be so if the itch were out of reach in the middle of my back – more activity would involve me finding a back scratcher. Furthermore, thinking again about my itch, I remember that, as a boy, I was told to desist from scratching myself. Underlining again how rules from other people can dominate life.

Thinking and feeling

> I am functioning well in my world
> when my ideas and affect are
> congruent and mutually supporting
> of one intention.

As discussed above, it appears at first sight that affective manifestations occur only at stage <3>. This is an exaggeration although true for the most dramatic manifestations. Affect expression begins there and can continue until stage <6> when it may develop and change as satisfaction is expressed at stage <7>. An "afterglow" may occur on withdrawal at stage <8>.

The "afterglow" may include residues of sensory impressions, including "anti-" effects as happens on going to a quiet place form a very noisy place, the ears are said to "ring." On leaving the strong perfume of a soap in the shower the outer room may temporarily appear to smell dank. Such "anti-effects" can be considered when an angry person suddenly quietens.

Other expressions may occur: a carpenter may be pleased enough as he starts his work <3> → <5> and then continue to express his happiness by whistling or singing <6>. Sighing does not always indicate sadness: my grandaughter (at 5 years old) smiles and sighs as she gets busy with something and often again as she finishes. She appears to be preoccupied and happy. An outside observer has no way of knowing exactly what affective expression means: it is necessary to check out. It would be erroneous to always interpret a terminating sigh as sadness.

> The only true cry
> the cry from the heart
> accompanies orgasm or parturition
> the harbingers of life.
>
> All other cries
> sadness, laughter, anger
> are cries of anguish
> of loneliness and death
> Be with *me*!

Multiple figures

The natural state for a healthy person is to have many figures operating within a small time frame, some of them intermittent and occurring simultaneously or consecutively. It is the unhealthy person whose vision narrows down well below Miller's famous 7 ± 2 items and indeed, in severe self-inflicted frustration may be one vanishing to zero in catatonia.

Thus the psychologically unhealthy person needs to concentrate on developing awareness in order to become healthy. This can be seen with a child who is concentrating, for example on writing. One foot may move and rub the other one. She may be chewing a toffee and there is music and probably the TV in the background. And she proceeds without impediment with her writing – until mother interrupts. The child knows no way of dealing with mother so she bursts with temper.

The one-figure-at-a-time rule is one of F. Perls' myths although, possibly, it suited him for himself.

It is sometimes not clear if figures are operating simultaneously or if the person is hopping from figure to figure, using each to interrupt the other(s).

Simultaneous figures

The relationship of sub-figures within a major activity figure can be very complicated. I have seen the controls of a large aeroplane – the pilot has to be aware of a mass of data while concentrating on one system at a time, concentrating by turns and exploring options. While he is moderately relaxed he is talking to his co-pilot from time to time – probably about football results.

The above exposition of figure stages worked on the naive assumption that figures developed and reverted one at a time like waves on the sea following one another on a beach. This is not true, of course. Figures overlap and interrupt one another as when talking while eating. This morning in the shower I was scrubbing my back while singing, tapping a foot and peeing – four figures simultaneously. F. Perls showed how useful this simultaneity is when he taught awareness to us therapists in a story that is now a cliché; a man is saying that he loves his mother while his foot is vigorously kicking out. "What" says Perls, "is your foot saying about your mother?"[54]

Sequences of figures

A figure may circuit in on itself. A man hammering a series of nails into wood does not have a separate figure for each embedded nail. He picks up the first nail <4>, hammers it in <5>, picks up the second nail, <4> again, hammers it in <5> again, and so on until all the nails are in. Then comes satisfaction <6> and the end of the job <7>, <8> and <0'>. The only concern here is that the inner cycling be uninterrupted, or if it is, that the carpenter returns to the point where he left off, after his tea break, for example. There will be many more illustrations of re-cycling and adventitious circuits in chapters to come.

Interruptions may occur with no deleterious effect provided that return to the interruption point occurs with resumption of movement in the figure A machine, the telephone is the most usual imperious interrupter in our times. Interruptions are not always disadvantageous and I list kiltered ones as –

• Something else occurs: all is OK with completion by return to the interrupted figure.

• Satisfaction may not be possible for reasons beyond the participants control. Thus in the desert I may be thirsty with no possibility of obtaining potable water.

• Satisfaction may not be possible if a fierce rogue dominates and does not allow. He may control the water supply.

Having established the validity of the figure stages I want to demonstrate the utility of using the system in the reality of the therapy room and elsewhere, although the pathological aspects will be held over for the next Chapter. First I want to examine one or two definitive propositions.

A non-organismic figure, as emphasized above, is contaminated by memories of irrelevant matter from previously failed figures

Two people

Each healthy person associates and engages with another and is active in his or her world by producing organismic figures. If the pair are in co-operative activity together their most comfortable state will occur if their figures are synchronized as illustrated in Chart 2·8.

This situation between two people can be subject of examination for incongruencies, rather similarly to the matching or lack of matching of the permutations of the Parent, Adult and Child ego-states in Berne's[56] analysis of transactions. A man is sawing a branch from a tree and his son holds a rope to control the sawn off branch. Both are at stage <5> but if the son's attention wanders he loses engagement as he wants to do something else, for example go for a pee, stage <3> of another figure. Father is likely to get angry because the branch will go out of control.

Although this style of relationship between two people would be a suitable subject for extensive examination in the next chapter this will not be done in this book.

Efflorescent figures

A healthy figure may branch out and produce sub-figures. An example will clarify my meaning here. With florid satisfaction the person may be so happy that a special satisfaction figure occurs, the person may dance around and sing in a way characteristic of that person. That this is a new figure can be seen by analysis –

<1>   There is a need to express delight
<2>   This need is recognized
<3>   There is a surge of a great, spontaneous feeling of joy
<4>   Engagement with self and everybody around occurs
<5>   The active expression with dancing and singing
<6>   The energy decreases
<7>   "Wow! Did I enjoy that?"
<8>   Withdrawal ready to get on with something completely different.

And similarly for sub-figures attendant on activity such a working, carpentry, for example. Again, these sub-figures are aspects of the character of the person – some folk never demonstrate satisfaction, some do. Some folk are hesitant as emotion bubbles up at stage <3>, others wallow in it. An ebullient healthy person may display all possible sub-figures in a burst of activity. An unhealthy person is unlikely to present anything.

Figure-stage analysis

The two sources of figures

As set out in Chapter 1, figures may arise prompted by needs originating from modern, cultural and environmentaly determined sources or from genetic, stone age sources. There is no bother if these are congruent and of health promoting effect. Bother may occur with incongruence but this is the subject matter of Chapter 4.

Awareness, attention and options

Awareness of figure flow can lead to interference because the dialectical issues are –

$$Flow \leftrightarrow Concentration$$

where concentration involves decision, and interference in the flowing process. Perls emphasized the value of concentration in coming to greater knowledge of what was going on but also stressed that the healthy state was that of easy flow, not pushing, striving or trying. Perhaps this is another example of differing attitudes for the therapy room and for real life. Flow and concentration are both aspects of awareness. Thus the word *awareness* deserves more than italicization – decorous presentation, exaggeration, colourful borders, poetic variations, sonata form and musical accompaniment are appropriate.

Then there is the matter of being aware that two or more figures are beginning to rise simultaneously re-

need to make a choice. I have just finished breakfast and have an option list in my head for what to do next – read the newspaper, go into the garden, sort out some books, go shopping, pay some bills and write letters.

Having such a list can lead to tactical choosing by sorting the initial options list into a priorities list or at least a list with the most interesting items at the top. Then the lead subject can be dealt with and the rest put aside for attention later. Tactical choosing can be very important in dealing with sequences of figures with related content because certain items must precede others. In the game of snooker a red ball is potted with potting the next colour in mind and beyond that the next red ball.

Dreams and fantasies

The dream is the most spontaneous expression
of the existence of the human being
F. Perls[55]

Spontaneous, yes, but existence no. Most of dream material is unwanted mental dross being excreted. Only sometimes, particularly when the dream repeats, night by night, it presents an aspect of self that is seeking completion. However, in my experience, "dream analysis" is one of the ways of avoiding important issues.

During the sleeping, hypnogogic, hypnopompic and day dreaming states, mind-stuff leaks into consciousness. The form may be memories or fantasy, each muddled by a multiplicity of material and each of the origin in well adjusted self or in alienated self.

As Perls stressed, the process of assimilation occurs most effectively if the dream is developed by role play in the present tense and with enactment occurring as if in the group room. Without judgement, role play sorts out the material of the dream and allows the dreamer to own and digest material or reject it. The person generally accepts centred self and assimilates useful alienated parts into centred self.

Assimilation is not entirely a process of acceptance. Rejection is perhaps more important because much dream subject matter is unimportant, rejected excreta; mind-shit. Just as there is no value in ruminating over turds, there is no value in dealing with most dreams, as most people know who don't bother with them. The idea of symbolic interpretation by an analyst (anal–yst) is a time wasting though a pleasant enough game. Any symbols belong to the dreamer who alone can interpret them.

Bern woke one morning having had a dream in which he was at table with many people who were all eating though he had nothing. The awake Bern felt hungry and went to his kitchen. In the group he recognised that he was prepared to eat heartily while someone else had nothing. While enacting the food he decided he had only to get it for himself, which he was good at anyway. The dream evidently had little relationship to anything other than that he was hungry.

Brett dreamt that he had urinated in his bed. Awake he found that he had not done so, needed to pee anyway and went to the toilet. In group attempts to analyse in terms of being his bed as a receptacle of pee or being the pee

flowing into his environment lead to re-enforcement of his ideas about himself as a person who fully integrated with his environment and who could control himself and be maximally effective. Neither idea was a revelation of Satori or Ah-Ha degree.

These are evidently two examples of dreams of centred self origin. Here is a dream story, as told by Bryn, that is abruptly changed. "I am in a house on the floor wet with sweat and bleeding. A woman brings me water and as I wake I wonder who she is". I hypothesize that centred self is only interested in intact wholeness and would not be involved in the double menace, cold wet river or gun shots, which must originate in alien parts. Role play, suggested by **T**, started with: "I will fall in the river and drown" and "I will shoot you dead" characters. After some enactment these two joined forces as: "We will kill you" and a third aspect appeared as: "I want to live." More time passed and it became clear that the latter was centred self now consciously dealing with a suicidal tendency. This was something **Bryn** had not faced for some years and, indeed, said he had forgotten about. Over some weeks in group he realized that he had this continuing aspect of himself that tended to sabotage what he was doing. By attending to detail of such sabotage in the context of the general tendency to kill (himself or, as it transpired, others) he gained ability to be more successful in his endeavours.

In relating to **Bryn**, **F** was strongly tempted to suggest a Perls style approach: "Be the river," "Be the river bank," etc. Those moves might have been as effective in outcome as an alienated self versus centred self procedure.

Ailie had just heard that she had passed her examinations in medical school and was elated but reported a dream that disturbed her and had occurred on three consecutive nights – "I am in the examination room and totally confused. I don't know which question to answer first. I think about one question and answers to another come into my mind. I waken, fearful and in a cold sweat. That happened several times during each night, the dream recurring in different rooms, not only at school but at home.

Ailie's attempts to identify with parts of the dream led to hilarity. "I am the exam paper – yuck!" She soon lost interest in her dream and began to talk about her future as a doctor. **Ailie** reported some weeks later, in answer to **F**'s question, that the dream had not repeated. Her dreams had become centred on patients who attacked her physically. **F** concluded that the examined dreams were a residues of worries that had plagued **Ailie** before the examination. After the examination success the worries were remembered and being irrelevant were effectively excreted.

Dreams also relate to parts of the figure circuit. Certainly, if the dream is regarded as an incident struggling to achieve completion then the facilitator's task is to facilitate completion. The common falling dream is a good example here. Usually the dreamer wakes up mid-fall, this is a stage <5> event. So "what happens next?" asks for reconstruction of stage <5>. Most commonly a soft landing occurs. Core-self is caring – and an ending <6> and satisfaction is achieved <7>.

Another variant dream involves unfocused affect,

Chart 2·8. *A generalized chart depicting the circuits of stages of a freeflowing empathetic relationship between two people.*

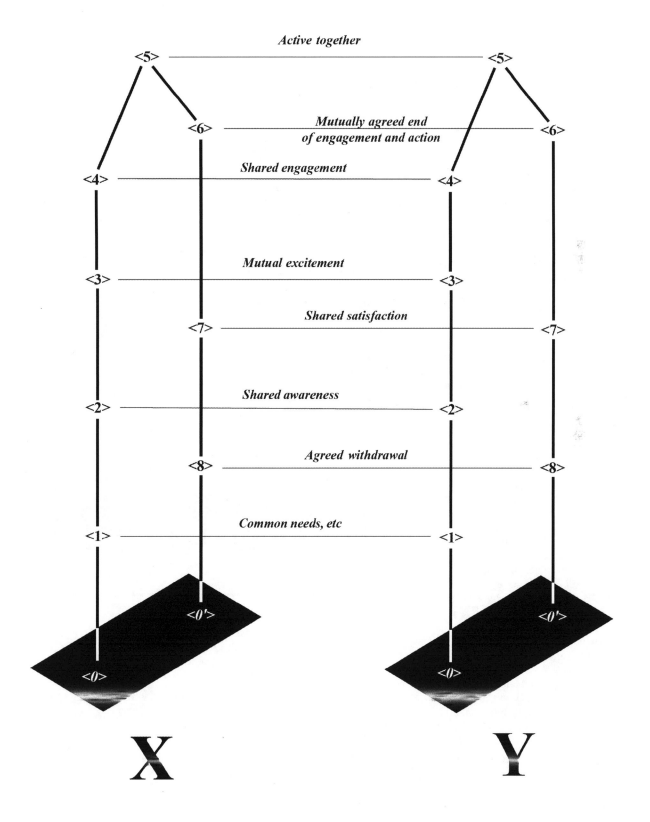

usually fear, stage <3>. Again the facilitating question is "what happens next?" Engagement <4> is made with something relevant to the dreamer followed by action <5> that concretises and develops the affect leading to satisfaction.

Resistance, assistance, avoidance and compliance

F. Perls[57] discussed emotional resistances and recognized three varieties, somatic, intellectual and emotional. In this present essay somatic matters are considered under the physiological heading, somewhat artificially separated from social aspects. Perls further clarified events as complete or incomplete; the worrier does not complete the action needed to express his aggression. Tactical choosing among activities, that is among figures, is often necessary and what may appear to be resistance and avoidance in Perls' terms may be simply delaying activity for comfort and some reasonable reason. Again, compliance, so indicative of confluence for Perls, may be a beneficial cooperative action. Judgement is as ever a possible pitfall and it is necessary to consider all possible factors including the polarity expressed as –

individual initiative ↔ team work

The rather simplified mechanical description set out above could lead to the impression that figures arise willy-nilly and that no personal choice operates. This is not so, of course, and apart from dealing with interruptions, it is possible to voluntarily change attention in a developing situation.

The point in having the information about the stages of figure development is to be able to use the information in real life, particularly in the therapy room and also in general education practice. Any further detailed discussion here would anticipate that of Chapter 3. Suffice it now to make the following general points.

Thought and affect are essential features of a healthy, self-nurturing figure. All forms of bodily sensation may occur and are indicators of mood while the figure is active. Thoughts are often discredited by gestalt practitioners and Perls particularly, as emphasized above, labelled intellectualizations and generalizations as "elephant shit," evidently indicating that he, personally, could not deal with such chatter. This was a grave mistake as such mental activities that go with avoiding what is going on also indicates the myths and rules that accompany the behaviour. Likewise, an affective outburst also provides avoidance of what could really happen. A small child who does not want to go out from home bursts into tears rather than say: "I don't want to go out." There are plenty of adults who use similar avoidance tactics.

OK people exchange information and freely express the full range of affect from sadness to ecstasy. They deal with points that arise and do not unnecessarily interrupt themselves or their companions. When there is a change of subject required: "Do you feel finished with that? I want to start a new tack."

Chart 2·9. The development of perception by stages.

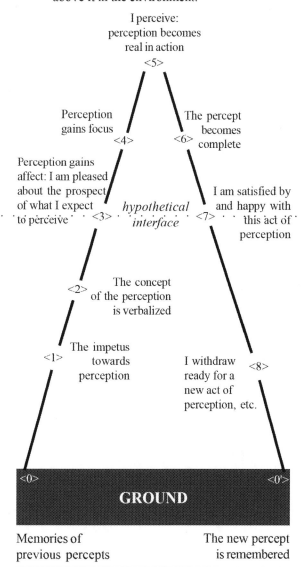

Perception develops internally below the hypothetical interface and externally above it in the environment.

I perceive:
perception becomes
real in action
<5>

Perception gains focus <4>

The percept becomes complete <6>

Perception gains affect: I am pleased about the prospect of what I expect to perceive <3>

hypothetical interface

I am satisfied by and happy with this act of perception <7>

The concept of the perception is verbalized <2>

The impetus towards perception <1>

I withdraw ready for a new act of perception, etc. <8>

<0>

GROUND

<0'>

Memories of previous percepts

The new percept is remembered

The person locked in alienated behaviour is effectively stunted and stranded. A primary disruption has occurred when the stunting is evident in interruption of the figure-stage processes described in this chapter. These interruptions are at the basic psychological level and will be extensively considered in the next chapter.

Protofigures as seeds

Discussion in this Chapter has neglected events in the ground, except for a note on page 19 concerning Lewin's "psychical tension"[38] as the origin of figures and Goldstein's "mean point"[41] as the place of return.

The nature of the protofigure in the ground is not directly accessible and must be inferred from the nature of the ultimate figure when extant; the figure carries many varieties of information from the ground, including, affect, memories, ideas and propensity for muscular movement. The protofigure supplies these though sometimes in rudimentary form. Thoughts begin to be realized at stage <2> of the figure (Frontispiece) and develop until they are probably most developed at stage <5> with consummation and satisfaction of the initial need, etc. Affect at stage <3> is an expression of excitement and, if highly energized, may disrupt the figure. I envisage that the subliminal source, as a protofigure, is a matrix with the potential to develop into a figure. To clarify with an example – if the initial source of the matrix was a highly traumatic incident, maybe in childhood, this matrix will carry the affect which will be expressed every time the initial incident is recalled, even though in vague approximation to its original format. If the earliest affect was devastating every reproduction of it will be a disruptive affective display.

In the first example in this Chapter the student depicted in Chart 2·5 frequently failed examinations, each failure making the affect from the examination failure protofigure stronger and thus making failure more certain. Perls' mother kicking group member evidently had a protofigure which expressed itself through his feet although his general figure was equable. When we come to Chapter 7 I will present information from 26 people and it will be clear that the depressed affect accompanying their figure expression compounds on itself with the consequence of making the person more and more miserable. The nature and effect of protofigures under those circumstances will be discussed after the data presentation in Chapter 7.

More sensitivity to figural processes

As Chapter 1 ended (page 30) I suggested that you could sensitize yourself and come to know more about your personal figures and gestalts. As this Chapter ends you can use the same –

Now I am aware …

mantra and become aware of the stages your figures go through as they evolve. You may realize that the stages for you are as I have set them out above, or as Perls, Zinker or Hall recognized them (table 2·1). Or you may come to unique conclusions of your own, and this can be to your advantage as you follow the developments in the Chapters to come and you can reinterpret my words into your own language.

The effectiveness of attention to gestalt figure stages was demonstrated by Wingfield[58] who dealt with the anxiety states of singers and other musicians.

Perception

To reiterate the statement on page 10 gestalt was originally a theory of perception. In dividing figure development processes into stages has the perception concept become lost? I am sure that perception theory is very much intact, well supported and reinforced and have summarized my ideas in Chart 2·9. We thus now have an account of stages of development of perception (Frontispiece[31]).

An earlier essay by the present author[59] was concerned with psychosomatic phenomena but was conceived before the importance of gestalt figure stage processes was recognized.

Adequate self-nurturing requires perception, attention, awareness and action as occurs in the healthy, normal every day life of usual folk It is time to get on with an account of inadequate self-nurturing and the interesting aspects of the consequences of psycho-wounds and how to heal them.

> The reward of a thing well done
> is to have done it.
> R.W. Emerson

Notes and references

1   Smuts, 1926, p. 106.
2   Lewin in Ellis[32], 1938, p. 285.
3   ibid, p. 292.
4   Reich, 1960, p. 115.
5   Perls, F. 1949b.
6   Perls, F. 1947, p. 26.
7   ibid.
8   Perls, F. 1948.
9   Kepner, 1987, p. 92, footnote.
10   Perls, *et al.* 1951
11   Kepner, 1987, pp. 433, 467 & 501.
12   We can only guess at what Perls and colleagues meant by "the given" which he refers to several times without elucidation: I guess it is what is going on in that place at that time.
13   I guess that the confusions within the 1951 descriptions are due to Goodman's failure to understand F. Perls' earlier formulations. One of the problems arises from their use of highly generalized descriptions without clarifying examples.
14   Hall, 1976, pp. 53 → 57.
15   Zinker, 1977, pp. 96 → 113.
16   Kepner, 1987, p. 91.
17   Polster and Polster, 1973.
18   Tillett, 1984.
19   Clarkson, 1989, p. 29. Wingfield (1999) used this model[58].
20   Parlett and Page, 1990, pp. 175 → 198.
21   Melnick and Nevis, 1992.
22   Carlock, Glaus and Shaw, 1992, pp. 191 → 237.
23   Rose, A., 1990, p. 19.
24   Wheeler, 1991, pp. 89 → 90.

25   Mitchell, 1996, pp. 23 → 43.

26   Schub, 1993, 79 → 111.

27   Perls, F. 1990.

28   Clarkson, 1989.

29   Ockham, 1646.

30   Perls, L. 1994, p 9.

31   When I have been teaching, using these chart forms, some students giggled: "how phallic!." Others have remarked on the hollowness, on yoni character. You, my reader, can interpret the chart as a metaphor in whatever way you fancy; perhaps with earthy, spiritual energy characteristic of a dolmen.

32   Ellis, W.D., 1938.

32   Lewin in Ellis, 1937, p. 284.

33   Lewin in Ash, 1995, p. 275.

34   Wertheimer see Ellis, 1937, p. 88.

35   Melnick and Nevis, 1992.

36   Figure-stages, as was written in the text, will be represented as <X>.

37   Ellis in a footnote (p. 283) emphasizes that Lewin is referring to mind matters and not to paranormal phenomena.

38   Lewin in Ellis, 1937, p. 290.

39   Wertheimer, see Ash, 1995, p. 281.

40   Perls, *et al.*, 1951, pp. 321 & 324.

41   Lewin in Ellis, 1937, p. 287.

42   The account of Zinker[15] and colleagues is based on a consideration that contact, here termed engagement, is the important stage and that action precedes it. I support the contrary, as expressed by Hall,[14] because there is usually comparatively little action during engagement. If **A** wants to talk to **B**, the talking is the need satisfying action <5> and moving towards **B** is preliminary to it <4>. The motivating energy begins to arise at stage <3>. There is, of course energy involved during stages <3> and <7> but the important, key, need satisfying energy occurs at stage <5>.
The nomenclature problem is solved by reference to the Concise Oxford English Dictionary. Contact is touching with no sign of activity whereas engagement is an activity.

43   Perls, et al., 1951.

44   Perls, F. 1975, p. 22

45   Miller, 1956.

46   Perls, *et al.*, 1951, p. 4.

47   Lewin in Ellis 1937, p. 288.

48   ibid, p. 292.

49   Simkin, 1994, p. 125.

50   Kepner, 1987.

51   Lewin in Ellis, 1937, p. 288.

52   Perls, *et al.*, 1951.

53   Maslow, 1962.

54   Perls, F. 1949a.

55   Perls, F. 1969, p. 71.

56   Berne, 1964 & 1973.

57   Perls, F. 1947, pp. 174 → 181.

58   Wingfield, 1999.

59   Edwards, 1985, p. 149.

—=(☆)=—

Part 2

Classical Gestalt Processes in Lack of Health

Mating with someone who is unhealthy
would have posed a number of adaptive risks
for our ancestors.
Buss, D. M.,2008, *Evolutionary Psychology*, p 122.

The classical gestalt processes of concern in this part of this book are set piece, reactive events which were referred to by S. and A. Freud and F. Perls as defence mechanisms since it was evident that, in the pathology situation, the person was defending self against the occurrence of a trauma, however minor of impact. Such defence was sometimes of doubtful benefit as the defence mechanism could itself entail alternative traumas.

These processes were also termed resistances by F. Perls since they occurred as resistances to a primary trauma. Wheeler (1991) pointed out that these events were not always of detrimental effect so that there were advantages in seeing these phenomena in a broader context. To aid this process he referred to contact functions which I modify to engagement styles[1] since we are usually concerned with the encounter of two or more people.

The engagement styles are subject of errors of development which can be examined on three logical levels[2] –

Gestalt – primary interruptions of the smooth flow of the figure stages, described in Chapter 2, which will be discussed in Chapter 3

Psychological – secondary errors when recognizable set piece events occur, including projection, introjection, etc., discussed in Chapter 4

Social – tertiary occurrence of heinous anti-social error events, including shame, guilt, etc., discussed in Chapter 5.

To illustrate the inter-relationship of these processes consider the introject "Beware of the Dog!" It may cause failure at stage <2>, preventing progress to <3>, because cold fear occurs. Then, as a beginning of processes of assimilation of the introject, the person says: "I should be ashamed of myself!" or words of that effect.

There will also be accounts of problems of behaviour of physiological origin in Chapter 6 including hyperventilation and anxiety, alimentary and food intolerance, secondary effects of endocrine pathology and stress.

Lack of health is a challenge to the therapist whose profession is to facilitate curative processes. Many challenges will be discussed and among them –

the value of doing as do the neurologists, seeing the benefits of association and detriments of dissociation

differentiating the influences of genetic make up from those of social and environmental factors.

References

1   See note 5 to Chapter 4, page 81.
2   Logical Level theory was presented by Bertrand Russell in his *Principia Mathematica*.

Chapter 3

Primary Failures of Interpersonal Engagement

Before us stands yesterday
Ted Hughes

The experience of Chapter 2 was entirely of association phenomena. As the gestalt figure developed the stages followed naturally with no impediment. Awareness of the figure was followed by realising that it had energy associated with it, was moving and/or changing, that there were options for development and that initial needs, wants and desires were satisfied. As all was well there were feeling of satisfaction expressed before going on to the next gestalt figure.

In this chapter we will be dealing with problems related to disruption processes, to dissociation from need satisfying activities.

As an introduction to generalizing about the run of gestalt figure stages in lack of health, here are a few examples .

Fifi was in her eighties and worried about shopping. She could go to a local shop but it was expensive. She could ask her daughter to go to town and shop for her. She could go to town with her daughter but had difficulty getting in and out of the car. She could ask a neighbour to help but didn't want to "impose on her."

Her need was for food, but she stopped herself getting things done by being unable to choose among options. She thus failed to prepare for going out and did not get her shopping done.

Jon had a high temperature, probably due to onset of flu'. He was aware that he needed water to drink but could not summon the energy to get out of bed and get it.

He knew his need, and did not move himself to begin to obtain satisfaction.

> The current central problem
> for physicists
> is not concern
> with the speed of light
> but with
> the speed of dark.

> All this gestalt stuff is OK.
> What about my ancestors
> of back to a million years ago,
> who interfere with me now
> by the action of my
> genes?

Matt was a carpenter who, one fine day, made a table. He spent days polishing it – "I must see my face in it!"

He was stuck unable to go on and say to his children: "I've finished. Look at this lovely table!"

Bram was talking about his work – apparently a unit. He then said he couldn't leave it alone and takes it home with him. The work unit is incomplete, aspects of it drift into his home life which is again incomplete. Two messed up gestalts. The remedy is simple. The facilitator encourages him to run a visualisation in which he slowly leaves his workplace, exiting by his usual door and wilfully leaving all aspects of work there, at the door. On returning to work by the same door, next day, he can, in fantasy, pick up work affairs again. By being slow and thorough **Bram** leaves the entire work gestalt where he wants to leave it, at work, and it does not travel home with him.

Bruce arrived and the facilitator watched him carefully. He was restless in his behaviour, his eyes didn't settle, he fidgeted as he sat in the chair. His words were not homogenous with one another. There was a sense of talking about bits and pieces rather than a whole unit. Anyway, he was not happy, he had come to therapy about something!

Bruce told the group how his Mother did not feed him properly and how he came to hate her. Two poor stage <5>s. The facilitator reminds him: "two weeks ago you praised her in glowing, loving terms. What is going on for you concerning your mother?" Evidently some OK stage <5>s. **Bruce's**

mother protofigure is clearly not homogenous. The process of healing will be one of resolving contradictions.

Meanwhile, which comes up most readily for him, loving or hating. **Bruce** can come to know, if he wants to, that he can love his mother for her good points and dislike her bad points.

I am now setting out a discussion of the consequences of interruptions in numerical order by using the gestalt figure stages from the last chapter. It may be clearer for the reader to start with stage <3> and proceed via <4> to <5> o r proceed to <2> and then <1>.

Interrupted figures, distorted gestalts and pathology

In a pile of white laundry
one red thread is enough to indicate
the location of the scarlet shirt.

Variation of function

Resuming the discussion of personal functioning from page 15, of the realization or otherwise of intention the categories of abnormality, gain, loss and/or change of function can be elaborated with examples in relation to the stages of figure development as summarized in the Frontispiece. These categories are another way of examining other features and all exemplify wounding by self-chosen frustrations. The following lists of examples are not sequences of stages.

Gain of function –
<1>   reaction to fantasy appetites; e.g., on seeing appetizing food on TV.
<2>   over-reaction to appetite, e.g., **Caleb** moaned continuously about a draft to the exclusion of all else that was going on around him
<3>   hyperactivity, e.g., after the ingestion of caffeine, etc.
<4>   overwhelming engagement, e.g., hugging to the point of nearly asphyxiating the recipient
<5>   Over energetic activity, e.g., making sure that one is seen by the boss to be a good worker
<6> & <7> overdone exclamations of satisfaction, e.g., again forcing a good impression on other people.

Loss of function –
<1>   no or diminished appetite, etc.
<2>   no or diminished awareness of curiosities, interests, etc.
<3>   no or diminished excitement, affect and energy, e.g., lethargy
<4>   no or diminished engagement with anything, e.g., no interest in other people
<5>   no or diminished activity, e.g., cooking is necessary

but "I can't be bothered"
<6> & <7>   no or diminished awareness of satisfaction as an activity becomes completed

Change of function –
<1>   mistaken appetite or interest, e.g., a child says he is hungry when he is tired
<2>   inappropriate awareness, e.g., on entering a peaceful forest one becomes afraid
<3>   socially unsuitable affect and energy, e.g., laughing loudly at a funeral
<4>   engagement with the wrong person or activity, e.g. blaming an innocent person
<5>   inappropriate activity, e.g., knowing that it is important to pay bills but reads books instead
<6> & <7>   inept expressions of satisfaction, e.g., effusive declarations of gratitude when little has actually occurred.

This analysis is simplistic, of course, since a person may display two or all three categories simultaneously. Thus **Carl** went hunting rabbits and met a wolf. So that no one should see his fear he climbed a tree and began to pick fruit. He rapidly increased his fear, then lost it as he gained height in the tree. Of course the wolf lingered below but that would be another story.

Functioning is still a question of completion or interruption of gestalts. Instead of the concept of loss of engagement with a figure stage we now consider the effects of over- or under-emphasis or change and avoidance of the original intention with respect to need or want. Thus a child may become so excited at the prospect of doing something that she become incapable of action, typically she vomits. A boy may convert free flowing joyous excitement into fear and embarrassment.

Resuscitation of interrupted figure development

Where was I when I was so
rudely, crudely and wantonly
interrupted?

In the following I will consider aspects of interruptions of figure development, disruptions of engagement and action in satisfying needs, etc., and what can be done to re-establish kilter and self-nurturing. The key activity is discovery of what is really required to have happen next.

As I will emphasize, people have selective wounding activities; they do not interrupt some types of figure and do interrupt others. They habitually interrupt at one particular stage of the circuit that becomes characteristic of the person. Attention and "healing" at that stage for one particular example then has a general beneficial effect whenever that stage occurs.

The unkiltered person is unaware of many aspects of his or her environment that would be necessary for effective, self-nurturing, happy life styles. The basic therapeutic manoeuvre is encouragement of development of awareness (pages 22 & 172), which then facilitates figure completion.

We will now look at the stages of figure evolution and discover the characteristic alternative behaviours as each is not realized because of interruption, avoidance of consequences, or resistance to going on to the next stage. Medical people could talk of pathology in most of these circumstances. Then we will generalize and particularize about these interruption behaviours that are the heart of problems.

Each sub-title is followed by the number of the page where discussion of the healthy, uninterrupted state occurred.

Before stage <1>; Absence of sensory presentiment of need for self-nurture, etc. (page 39)

Beware delusions of safety
James Baldwin

If neither mind nor body sensation or appetite occurs there is neither gestalt, ground nor figure. Sensations are completely suppressed, representing a zombie state of being, illustrated by **Alma, A**. The facilitator could confuse this state with the impasse but the latter follows a bout of dealing with confusion. Likewise the meditation state has about it a feeling of peace that the shiftless, suppressed animation of the dull, apathetic person does not have.

A went to her general practitioner doctor and was told that she was anorexic. She agreed that she never felt hungry and only eat anything when other people told her to. She dissociated herself from her basic need for food.

F asked: 'What do you gain by not eating?' No answer.
F: 'What sort of things interest you?' No answer.
F: 'Are you curious about anything at all?'
A: "I wish you wouldn't nag at me!"
F: 'What should I do instead?'
A: "Just talk ordinary with no questions."
Liberation! **Alma** made an engagement figure with **F**.

Goncharov created Oblomov and depicted him as spending his life in bed, served by a peasant, and only slightly interested in his external world when the ice broke on the Volga in spring.

Before stage <2>; Blocked awareness of need for self-nurture, etc

(page 40)

The road to hell
is paved with good intentions

Stage <2>, like stage <1>, can only occur in present time in a particular place.

The person who is uneasily aware of bodily sensations and is thinking about something else is blocked off before stage <2>. I remember as a teenage student concentrating on solving mathematical problems and being only vaguely aware that I was hungry and certainly not concentrating on it.

Hall (page 37) laid emphasis on lack of organismic flow as characterized by dullness of awareness, ambiguity and the tendency to have cognitive processes focused in the past or future. I add to these dullness of emotional recognition and expression.

Chad: "I'm very cold and I think you ought to keep this room warmer." After some facilitation of process **Chad** realised that he had been slowly getting colder over about half an hour and really did not recognize it until he felt very uncomfortable. The group joined him in his experience of cold outside the house (it was winter) and in hot air from an electric blower fire. Later **Chad** remembered that his mother always shouted at him if he said he was cold or hot saying that he was unbearably sensitive. He shouted back, eventually in group, that he was ordinarily sensitive and that he was going to look after himself. In further experiments, in company with the whole group, he learned to feel his heart beat in his chest and to notice a feather tickle on his neck.

Anita sat trembling for about five minutes before suddenly erupting into anger. **Claud** developed a runny nose and sniffled for some time before his eyes glistened with pre-tears. Only some weeks later did **Claud** actually sob and cry wet tears that expressed the feelings that he had dissociated from.

Re-establishment of recognition of awareness of sensation. F. Perls and others have dealt with the basic manoeuvres for re-sensitizing a person to recognition of appetites, awareness of personal needs and the facilitation of curiosity. The method consists of concentrating attention while talking about it; "Now I am aware that I am thirsty and I need a drink." Extensive facilitatory exercises are set out by Enright,[1] Stevens,[2] and Paige[3] and I will only elaborate further in relationship to specific examples that turn up in later pages.

Introjective processes need to be dealt with here. **Clive** has a rule which he expressed in the form "I don't count and I always look after other people first." As a projective process; **April** said "I should not need to bother with food. Someone else should always bring me what I need, on the hour." The retroflective expression can take form as when **Ava** was watching starving refugees on the television and turned "You should not feel hungry" into "I feel hungry – though it is really nothing to do with me."

F. Perls made much of the ability of some people to be unable to recognize certain, specific sensations, to scotomatize them as he said, and yet recognize other sensations[4]. **Beah** was described by her G.P. as anorexic. She drank lots of water in a day to satisfy her thirst but she said she never felt hungry[5].

Before stage <3>; Interruption before development of excitement, energy and affect (page 40)

Mumblers know
that all other people are
deaf or inattentive.

Stage <3>, unlike stage <2>, starts in present time at a particular place but may soon be subject of wandering in space and time as memories of the past and anticipations of the future come up. These may reinforce or disrupt excitement and affect development. This stage is, for many people, the particularly vulnerable one where the decision is made to choose among options, to satisfy the original need, etc., or abort. The latter is a wandering away from original intentions. A dissociated, out of health figure state has been generated.

A lethargic person sits and contemplates his need for companionship, saying to himself: "Shall I visit **X** or **Y**, ask **Z** over or just go for a walk. I could go dancing tonight, etc., etc." Dissociation by procrastination becomes a substitute for self-nourishment.

Hall (page 39) contrasted organismic excitement with anxiety and panic states of the an-organismic person. He emphasized the importance of Perls' observation on breathing and counterposed the steady, effortless, full, deep, economical breathing of health to the restricted, upper-thoracic breathing of the anxious person. Avoidance of real affect leads to experience of a different affect that recurs each time this situation arises and I call this the familiar unhappy feeling of the pseudo-figure which is really a new figure, efflorescent like the euphoric figure. For an aware person this event signals that something is wrong and a choice to do something different.

Here the mechanism of dealing with the interruption follows the previous pattern. A choice is available, realise what has happened (awareness), and get back to the excitement that had been blocked and move on to engagement or get into the familiar unhappy feeling and degenerate any excitement energy and affect that have begun to develop. This situation is represented on Chart 3·1.

As illustrated above concerning stage <2> most children are shouted down when they get excited about something and the parental injunction to "good behaviour" may last a lifetime and may be reworked spontaneously or in therapy.

Several groups of students, at first in pairs, then all together, shared their answers to the question: "What message did I get, as a child, from the adults around me, when I became elated with excitement, exhilarated, transported with joy, 'super-euphoric,' wild with joy, over the top with happiness and/or ecstatic?" Many of these were the student's own terms and my intention was to focus their attention on happy excitement.

Three had a message like: "Stop jumping around. You know it will only end in tears." "You will end up in the vale of tears if you behave like that." And by reports it did

because a good child obeys mother's instructions. Other messages were –

- "You will become ill."
- Mother became very angry, verbally squashed the child and forbade any form of excitement, especially happiness.
- A Nun became violent and beat the child of about 6 years.
- "You will get an asthma attack" – and she did.
- "You will harm yourself and damage the furniture."
- "You will become depressed"
- "Just you go away somewhere, quietly by yourself"
- "As a punishment you will go to bed without your supper."

Each of these of a great variety of messages, received by healthy citizens, initiated years of realised expectations; scolding, anger and beatings, asthma, illness, isolation and tears for self. These were wise students and claimed that they were not at the time of the experiment, subject to the same reactions. I believe them because I believe in self-healing.

On another occasion students shared what they did with anger. All said that they could not be angry or bad tempered when they were small because they were punished if they did. Other observations were –

- "I was anxious about my heavy heart beat and expected to become ill with heart disease."
- "I got very bored while trying to think of the best thing to do."
- "I wanted to be angry but couldn't and this got on my nerves so that I got tired and sleepy."
- "I felt all the muscles in my chest tighten so that I couldn't breath." Another student had a similar experience.
- "Everything was too powerful for me and I switched off. I got very vague about what was happening and got dizzy as I held my breath."
- "I felt like hurting someone but knew that I shouldn't do that. I could only hurt myself and that was O.K. Once I burnt my finger a little on the gas as punishment for wanting to scream at my father."

These were people from the general population who were interested in the psychology of counselling skills. They were adding adult interpretation to childhood experiences so these are very revealing statements about childhood learning. Every person had something deleterious to say.

Dealing with the familiar unhappy feeling of pseudo-figures. People usually don't realise that they have a "favourite," familiar bad feeling which they take on themselves at times of harassment and stress. So one of the valuable awareness moves is to help the experimenter to discover and really get to know his or her familiar unhappy feeling. An effective way to do this is to deal with a direct question of the form "What do I do to myself when something goes wrong for me?" Alternatively the fantasy trip set out by the Gouldings[6] is valuable. These are, Berne's "racket" feelings. It is the affective residue of a dead figure.

If I can't face the affect at stage <3> – or ideas about them – I take myself off somewhere that appears to be safer, even if that is only a fantasy. And I need to reformulate negative

Chart 3·1. Interaction of two incongruent people; interruption as feelings rise and before being really excited.

boundary / interface

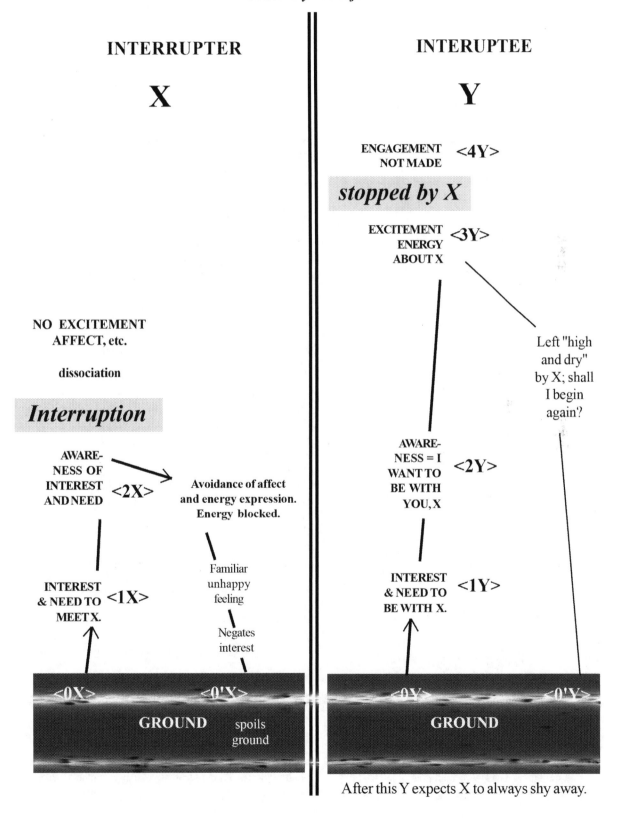

INTERRUPTER

X

INTERUPTEE

Y

ENGAGEMENT
NOT MADE   <4Y>

stopped by X

EXCITEMENT
ENERGY   <3Y>
ABOUT X

Left "high
and dry"
by X; shall
I begin
again?

NO EXCITEMENT
AFFECT, etc.

dissociation

Interruption

AWARE-
NESS OF
INTEREST   Avoidance of affect
AND NEED   <2X>   and energy expression.
                  Energy blocked.

AWARE-
NESS = I
WANT TO   <2Y>
BE WITH
YOU, X

Familiar
unhappy
feeling

INTEREST
& NEED TO <1X>
MEET X.

Negates
interest

INTEREST
& NEED TO <1Y>
BE WITH X.

<0X>   <0'Y>   <0Y>   <0'Y>

GROUND   spoils   GROUND
         ground

After this Y expects X to always shy away.

injunctions (introjects) and create positive alternatives.

When groups of students considered this question they were asked: "What is your familiar unhappy feeling?" and "What do you do about it?" All students readily produced answers to both questions. To the first question came, with number of students as {X}, distracted and bored {12}, energyless and depressed {11}, angry with myself {6}, angry with someone else and blaming them {4}, sad {2} and giggly {1}. (A degree of liberal interpretation has been used in categorizing). Answers to the second question about response were: turning on music, dancing "around a bit," listening to TV, radio, tape, or CD contrasted with turning on fantasy music by singing or whistling or making noiseless music in the head. Also "imagining I am somewhere else" particularly in the country {2}. Each of these healthy people had awareness and a way of dealing with unwanted situations which is self-therapy.

Experiments carried out by students showed that they had ways of controlling unwanted emotions, as is necessary in an adult, civilized society. **Clem** was strongly criticized by his boss at work and had difficulty controlling both tears and anger. At home, at bed time, he both cried and thumped his pillow. Discussion by the students led to recognition that the self-defeating emotion was generally dealt with by returning back down the figure. Instead of proceeding to action <5> each person retreated from engagement <4> to awareness of affect and energy <3>.

After considering the activities of the students it became clear that the unkiltered people do not have the ability to turn off their familiar unhappy feelings and thus return to an organismic state. Emotions wander willy-nilly. In health the developing figure is partially characterized by its affect. In lack of health inappropriate emotions are exhibited. **Cody** was a reasonably good footballer but if challenged when controlling the ball he produced flashes of anger which usually resulted in the presentation of a yellow card by the referee.

The general conclusion was that safety was gained by proceeding backwards in the figure in the manner related above. Instead of excitement and energy rising and being ridden by affect, affect was diminished by returning to generalized excitement which again subsided while a new, analytical figure took over: "What's going on here?" .

So how about replacing the familiar unhappy feeling with a familiar happy feeling?

Re-establishment of the development of excitement, affect and animation. The basic procedure is as described above, for sensation, stage <2>.

Beryl: "I always end up unhappy. I can be having a good time then something happens and I start crying."

F: "Remember the last time you did this to yourself. Tell us about it as if it was happening now."

Beryl: "My sister was getting married and I went to her party. I was having a good time with the bridegrooms brother when I knew I was going to cry. I ran to the ladies room where an Aunt comforted me while I blubbered. I told her he had said something nasty to me but I don't think he did."

On the third time of slowly going over this scene **Beryl** remembered many very early times when her mother shouted at her: "Don't get so excited. You know you always end up in tears." **Beryl** laughed: "She told me to cry and I did. Now I am not going to cry like that anymore." **Beryl** followed **F**'s suggestion and visited each group member, getting more and more excited and laughing gaily, telling in many different ways how she was going to enjoy herself and be excited and continue to enjoy herself until she was tired and happy with exhaustion.

A similar scenario involved **Bess** who had a message from mother about getting excited and then getting sick, meaning needing to vomit. On this occasion **F** fancied a different way of ending with contented excitement. The whole group played "Nuts in May" and at first stopped each time **Dee** said she felt ill. Then started up again when **Dee** was ready until **Dee** was whole heartedly immersed in rowdy joy.

Biddy said: "I should not need to exert myself – I have servants." **Cosmo** said: "I don't need to be angry about business matters – I have a manager to take care of that."

The basic procedure is as described above, for sensation, and in fact also applies at each subsequent stage of the circuit of figure development and involution.

Pseudo-euphoria. One of the interruption processes that I regularly dealt with for myself, many years ago, was a radio voice from across the corridor. One Sunday morning I heard an enthusiastic woman's voice eagerly asserting that it was necessary "to perpetually debate the nature of the living Christ."

It struck me immediately that she was using interruption processes to prevent herself from engaging with ordinary reality and that these were deflective and similar to the interruption processes of anxiety, except that she was clearly experiencing joy. The interruption allowed her to distance herself from the inbuilt absurdity of the slogan she was chanting.

I provisionally designated her condition as pseudo-euphoria and generalized it as I saw it in other people. The process is used to hide real, underlying feelings and ideas about feelings, the simple ones, loneliness, sadness, fear, and anger, envy and other mixed emotions

The signs and symptoms of pseudo-euphoria include a manifestly untutored, ignorant attitude, and a countenance that is rubicund, mellow, caryatidic and lugubrious. It has an epidemiology like a yawn which may be "caught" from others and, for this reason, is much used by proselytising persons.

The treatment of pseudo-euphoria can be the same as for anxiety and panic –

STOP!
Ask: "What's going on?"
"Is this true for me?"
And do something completely different.

The state of pseudo-euphoria can be part of the character manifestation of "Manic-depressive" experimenters when in the mania phase much as anxiety is manifest in the "depressed" phase.

The phrase "what's going on?" was regularly used by one of my group members and I did not tutor him in a response to his own processes. My contribution was to point out that the use of this phrase provided a stop to the pseudo-euphoric process that he generally, gaily, presented and that this reaction could be made stronger by preliminary use of the word "stop." This group member was then in a higher energy state and found it easier to be aware and clear about his false emotions and about his intellectualizations. To raise the energy level is to move towards a curative engagement and action circuit and the true nature of the deflective process becomes obvious.

In summary, the use of pseudo-euphoria by proselytizing Christians and energetic acolytes of other religions or belief systems, using the tactics of whirling Dervish dancers, communal rituals to avoid confrontation with the nonsense of the ritual as well as other aspects of life. Religion in this space is truly an opiate. The rituals of valid religion and theology do not need to impose sophorifying pseudo-euphoria on adherents.

Before stage <4>; Interruption before engagement and mobilization of resources (page 41)

The anorganismic person may avoid engagement altogether or just fumble self-consciously. A particular example of interruption before stage <4> occurs when someone dissociates by trying, even trying hard, instead of doing, getting on with the necessary action.

When person **X** suddenly stops his movement into engagement with **Y**, **X** is aware of emotion he feels and does not want to express. **Y** picks up signs of the emotion anyway and wonders what is going on. Again, **X** senses this and becomes embarrassed, shy, diffident, self-conscious; he has his familiar unhappy feeling among many possible ways of expressing his discomfort.

If **Z** is courageous he can share his feeling: "I'm getting myself randy – let's do something completely different" and engagement then develops. Here **Z** has put himself back in engagement instead of going into his familiar unhappy feeling, out of engagement.

Re-establishment of engagement with the need satisfying action. Some people are afflicted by a dread of engagement with other people.; **Bunty** said: "I will be swallowed up." She needed to experiment to discover that she was, in the present time among a group of O.K. people, and not in danger of being swallowed up. **Cyril,** an ex-serviceman, was afraid of groups of people who would, he said, always try to control him, was afraid of dominating people. He valued his liberty. He learned in group that nobody was out to control him, least of all the facilitator. He was given some experience of controlling the group, marching them up and down in pairs.

Cyrus had an introjected rule that he must always be excessively polite so that he always inhibited interpersonal approaches.

Cyrus: "I'm not doing this – you do it for me."

Selective engagement was sometimes the effective process – **Dale** said; "I should not need to have anything to do with your engagement and action. These are menial tasks. I will read and enjoy my Latin texts – that is really being civilised." **Dale** was not concerned with the protestations of other group members. Why should he as he had servants to provide for his basic needs.

This is the stage where excitement and feelings are blocked from flowing into muscular (motoric F. Perls says) behaviour and the blocked affect appears as turmoil and "Psychic pain." We then pretend to do something else and play roles.

The next stage would involve action which can only occur in the now and at a particular place. Once started action continues in that continually evolving space and time unless disrupted. However, action requires mobilization of tools and ideas from the past and anticipation of the needs of the future.

Before stage <5>; Interruption as the action occurs (page 41)

Hit first
talk after.
Principle taught to British soldiers, 1992

A person can be teetering on the edge of engagement and not become active and associating. I watched a shy lad become excited and I presumed he was about to say "Hello" to a stranger. He made eye contact, reached out a hand and opened his mouth but then blushed and backed off while saying something incomprehensible and practically inaudible. He had made contact but without interpersonal action. Another victim of shyness, **Dane**, was able to pick up a pen, hover it over the paper and not write a word. **Dodd**, in group, was stretched out, flopped over a cushion and said "Well ...." While looking around, aware of the group's attention. Then "Later...." as he slumped down again.

Activity is often interrupted by someone else immediately before achieving the action and the kiltered person copes without bother. A typist who is interrupted by her boss needs to be able to carry on typing where she left off.

Before stage <6>; Failure to realise that action is complete and the need is met (page 42).

Fussing around often indicates that something is incomplete although it can also suggest that something needs to be started. If I enjoyed stage <5> and stayed there for a while I am not necessarily resisting going on to stage <6> and satisfaction. Here resistance would be the wrong descriptive word; happiness might fit better. It might only be right if I am so negative about myself that I can't stand the state of being satisfied with my activities.

Perfectionism is failure to realise that what has been done has been done effectively and adequately well.

F 'I'm O.K. with 95%. How about you? Can you tolerate 5% errors, that's 1 in 20 events not turning out exactly as you want?'

E "I don't like making any mistakes at all."

This form of failure occurs when a person has not completed the action necessary to satisfy the need, etc., but makes some kind of gesture which might convince another person that he had. Genet wrote many examples of this phenomenon. It was said of **Duke** that when he finished a cabinet making job that he got stuck with the polishing stage. He went on and on for days, seeking perfection.

However, **Dean** said, "I'm not into all this energetic assertive expression. Stop pushing me. I'm soon joining a Benedictine order and opt for a peaceful life."

* Doon's house was a muddle of objects, each in a wrong place. There were carpentary tools in the bathroom, left there when a job was done but not finished by replacing tools in the garage. Other people in the house were frustrated by being unable to find tools at the bench in the garage. Odd pots of paint and brushes could be tripped over in most rooms. Doon also cooked delicious food but left the kitchen in a muddle of uncleaned pots and pans. In terms of figure stages, .Doon was happy to proceed to stage <5> but not beyond where stage <7> would occur.

The therapy group can be a safe place for embarrassment to develop and be shared. Behind the shyness is a message the most common variant of which is "You think I'm foolish!" **X** can check around the group saying, member by member: "Do you think I'm foolish?" He discovers a variety of thoughts and feelings and no allusion to foolishness. He can also role play someone from his life, or a fictitious person, to test the reality of the foolishness message in that context. There is also the projective aspect. He can say: "I think you are foolish!" which can be explored.

As above in dealing with interruption of activity, re-enactment at a slow pace, perhaps two or three times, brings up memories of earlier interruptions to engagement that can be dealt with.

Helpful facilitating questions are –
- What do you gain by breaking off from activity?
- How can you remake active, creative engagement?

Visualization of successful engagement occasions from the past reinforce a concept of success and imaginative future projections can be very beneficial.

Breaking away from repetitive activities. The issues surrounding occurrence of inability to recognize that a need has been met arose in 1:1 work with **Doug** who presented with persistent, repetitive behaviours. He checked three or four times that he had shut his front door as he left his house. He washed his hands 6 or 7 times after doing a dirty job like rubbish disposal.

There are two elements here, repetition of the action itself and repeated checks that the action has been carried out. Two other aspects are –
- that his memory fails, he literally forgets that he had checked the door, that he had washed his hands already, etc.
- an injunction is at work and is of the form "do not complete what you are doing" or "I approve of what you are doing so keep on doing it."

In relationship to the latter he remembers that he rarely

completed what he was doing as a child. His mother or nanny would stop him and get him to do something else. This continued at school where, an example he gave, he would not have completed a drawing within the time span of the class.

He did not recall a specific instance of his mother, nanny or anyone else making an interruption; this could have come up again in a future therapy session.

On having his attention drawn to the action process circuit he recognized that, though he had completed his activity, stage <5>, he had no sense that his need had been met, stage <6>. He noticed that his tendency was to go back to the awareness stage, <2>, and start again towards engagement and action, thus being trapped in an ungrounded subsidiary circuit.

As **Doug** was leaving a session I noticed that he had no repetitions operating as he put on his scarf and buttoned up his overcoat and I shared this observation with him. It is not possible, at the time of writing to evaluate the long term effectiveness of such observations.

Before stage <7>; Lack of aware satisfaction and validation (page 42)

Probably the most common interruption of the circuit of events of figure consummation in our hurried and harried society is failure to feel satisfied and not feel positively good about completion of engagement and activity. "One must hurry on" thus dissociating from the pleasures of success and gratification.

Everyone benefits by attention to this:
F 'You can give thanks to your self for every job well done. Bask in the contentment of it all.'
This is self stroking, as the T.A. people call it. It is also beneficial to give strokes to the other people involved in an enterprise and to enjoy receiving strokes from the other people.

Hall[6] commented on the dissatisfaction and dysphoria of the an-organismic person who may feel a sensation of "indigestion" and confusion due to lack of proper assimilation of what has been going on. This person most usually had no sense of anything at all, or may interpret satisfaction as dissatisfaction by using lame cognitive nonsense.

Facilitating awareness of satisfaction and valuation. The activity is complete, I have a sense of satisfaction and do I value what I have done?[7] **Earl** had a rule operating: "I must not overwean myself, I am not that important and should get on with something else rather than praise myself[8]."

The benefit of a thorough evaluation of a success is that impressions are carried with memories into ground so that future figures, even if not so similar, are more likely also to be successful. There is another extreme, of course, involving overdoing self-evaluation. Self-glorification can have a detrimental effect on other people who become unco-operative and sick of it.

```
┌─────────────────────────────────────────────────┐
│                  First Fool                       │
│                                                   │
│  E    (After a long "silence") I'm afraid of      │
│       making a fool of myself.                    │
│                                                   │
│  F    What do you feel now?                       │
│                                                   │
│  E    Easy, actually. I feel lighter for having   │
│       said something.                             │
│                                                   │
│  F    And before feeling easy?                    │
│                                                   │
│  E    I felt very tense in my ribs – not          │
│       breathing adequately.                       │
│                                                   │
│  F    And now –?                                  │
│                                                   │
│  E    Breathing OK. No tenseness. (smiles)        │
│                                                   │
│  F    And before the rib tension –?               │
│                                                   │
│  E    That was when I chose to come into action.  │
│                                                   │
│  F    Into action, Eh! Do you often tense up      │
│       before action?                              │
│                                                   │
│  E    Often but not always.                       │
│                                                   │
│  F    When not?                                   │
│                                                   │
│  E    When I'm by myself. I tense up if someone   │
│       is watching you.                            │
│                                                   │
│  F    Watching you or me? (with hand gestures).   │
│                                                   │
│  E    Watching me.                                │
│                                                   │
│  F    How do you feel now? I – we are all –       │
│       watching you.                               │
│                                                   │
│  E    A little tense. (laughs with a deep breath).│
│       That's it; OK now.                          │
│                                                   │
│  F    So what will you do if you feel tenseness   │
│       in your rib cage?                           │
│                                                   │
│  E    Breath (short pause) and laugh (laughs).    │
│                                                   │
│  F    Ask around. Did anyone see you as being     │
│       foolish?                                    │
└─────────────────────────────────────────────────┘
```

Before stage <8>; Failure to ground (page 42)

Nothing succeeds like success.

Hall contrasts the fertile ground with the lack of reality of the barren ground, as I call it (page 37). This is the place of undifferentiated anguish, *angst*, to use the German word with existential connotations. This person hangs on oblivious of the fact that there will be no new figures to develop and appears to be stupid while nothing further happens. In health the ground is reached after fulfilment of all the steps of the circuit and the personal gains of the beneficial experience. The experience is remembered. The ground may be significantly altered, enriched in a material sense, as when a work of art or craft is created.

As described above, the ground is not as it was when the figure began to emerge, as Hall emphasized. The an-organismic person returns with a futile, familiar unhappy feeling (page 58). For him or her the harmful behaviours recycle, there is no rest, energy of contradictions locks in an impasse, the predominant feeling is of disintegration and lack

of balance and he or she may turn on self with verbal castigation and physical punishment. Self-nurturing has failed.

Persistent failure of figures results in return to ground in an unsatisfactory way, so that the ground then consists of a mess from which new figures emerge, already distorted. There is the expectation that such figures will not have adequate emotional or energetic back-up, that engagement will be bad – that action will be fruitless – "I never get on well with people" a self-made deleterious injunction or psycho-virus. A decision will be made – "I never was any good at making things." This is barren ground communicating, to over emphasize. It is a source of suffering and misery.

A false reason for not stopping may be indicated by words like: "I like you. I want to go on with what we are doing together." "If I stop he will give me something else to do." "You keep on going, Son." "Don't stop or I will feel lonely."

Facilitating grounding. Anyone who hangs around working out the satisfaction stage can be helped to appreciate moments of rest, peace and quiet, of re-establishing connection with the possibilities of effective affect and energy in forthcoming figures. The use of guided fantasies is particularly beneficial. Couè style affirmations[9] can be very effective at this stage: "**I am** an OK person and **I am** successful in what I set out to do!"

Before stage <0'>; The resting, centred place, ground (page 43)

Grounding is both the end and the beginning. Here the protofigures that developed arise as potentially the next figure. Alertness, liveliness, "next episode please": not meditation which is internally focused and not the impasse where transfixation by dilemmas occurs in uneasy stasis. Here also occurs out of awareness assimilation and learning from experience. Alternatively here is a novel stuck state where there is neither need, awareness, excitement, engagement, action, satisfaction nor gratification, mild catatonia[10] all associated with dud protofigures.

If there are no signs of confrontation with confusion, the impasse, the facilitator can guide the experimenter back into engagement with himself. "Now I am aware!"

Objectives and accidents; multiple figures

What I can't have
you shan't have
smash!

In the usual run of life the interruption process and running multiple figures is no problem. Times occur, though, when a figure is running and an interruption causes unforeseen change. My hands and the rest of my musculature are about to place a glass bowl on a bench in my kitchen. Thoughts suddenly intrude – it would be better on the shelf above the bench. The result is confusion and the bowl collides with the kettle and breaks. The way to avoid such a dilemma is

---

Problems with men

E I am anxious all the time. I don't know what to do about it. I've come here hopi"ng you will know what to do.

F I don't know what to do about your anxiety except help you to know what you want to do about it.

E [whining] I don't know what to do.

F So you know that something can be done.

E I suppose so. I wasn't always like this.

F When were you not like this?

E When I was in my teens I was happy enough.

F When did you change from being happy to how you are now?

E I married. It was OK to start with but we began to quarrel about things, I like things one way and he always wanted another. He used to hit me so I always agreed with him.

F I suggest that you dialogue with him. Put him on that chair.

E He left me in the flat with no income. He wrote to me from New York . I couldn't pay the rent and went to live with my mother. I've felt so miserable ever since.

F Did you write to him?

E No. I don't want anything to do with him any more.

F Is there anyone you do want to know better?

E Not really. I have a job. I see people. But I don't trust men any more.

F Do you trust me?

E [More an imated] Oh, you are different.

F The men you work with – do you trust them.

E Well yes but that's at work. I wouldn't live with one.

F Are they married?

E Yes, they talk about their wives.

F Do their wives trust them.

E I suppose so.

F So you could trust them, don't you know?

E Yes, of course, I could talk to them and keep house but how would I know if they would be quarrelsome?

F Put one of them on the chair and ask him.

E Would you always agree with what I wanted?

F What is his name?

E [Much more lively] He's my boss. I'll call him boss.

F Ask him your question?

E The question. Yes. What would you do if I wanted frilly curtains and you wanted plain?

F Be him.

B I'll not be bothered. Be frilly if you want.

F Put some more questions.
And she did, half-a-dozen of them until –

E I don't feel anxious about him.

F And the other men you know, would you be anxious around them?

E I suppose not, really. They are not like my husband.

F Let's stop there. Notice how you feel about men, this week, and we can talk about it next week.

---

to proceed slowly enough to enable adequate co-ordination. **Edwin** is about to tell **Dinah** that he loves her – she picks up the non-verbal cues – but he interrupts himself and asks her out to supper. She is not sure why she is confused.

Since excitement precedes emotion the essential move in dealing with unwanted, out-of-figure emotion is to get back to the feeling of excitement and to choose what happened next; the in-figure emotion usually offers itself immediately. Such Out-of-figure

emotion is feeble and the affected person appears in public as cold, indifferent, dispassionate, dull and very boring.

Useful leading suggestions and questions include –

- Be aware of what you do with your excitement.
- Go back to being excited.
- What can you do now with your excitement?

Formal awareness experiments help group members to an increased sensitivity to both emotional experience and to knowing other sensations in the body.

APPLICATIONS OF FIGURE STAGE ANALYSIS: GOING FORWARD TO CONSUMMATION

The point about knowing about gestalt figure stages
is to be able to use the knowledge
for the benefit of people who need to
complete incomplete activities.

Lithium

E1 here has been diagnosed by psychiatrists as manic-depressive and was being treated with lithium.

E1 [Smiles beneficently, looking around the group members] I'm the messiah. I have come again to save you all.

F Do I hear correctly that you are saying that you are the messiah?

E1 [Very softly] Yes. I have decided that I am the messiah, come again on earth.

F I like the idea of you being a messiah, a very kindly person wishing good will to all men.

E1 [Just sits silently, smiling radiantly at the group members by turn]

F Would you like to know what is going on for us, one by one?

E1 Oh, yes. That would be so kind. [Looks around as if to invite comments].

E2 [After a pause]. I don't know what to say, really. You think you are the messiah and that's OK with me except that I have my doubts.

E3 If you had said that in a pub I would have laughed at you. Ridiculous, you know. You're not even a christian.

E1 Oh but I am. I am the Christ come again, as was promised.

This kind of exchange continued during that group for a few more minutes and in subsequent groups, on and of, for four weeks. During that time **E1** became less and less confident about his messiah conviction and began to make excuses for his behaviour. Group members were very understanding and supportive of **E1**'s changes.

Toilet training

F: I fear that I will be swept into some orgy scene that I would hate.

F: Can you bring your attention back here. You told us last week that you had to be strong with yourself to enter this house and room. [no response]. What did you expect?

F: Nothing much; nothing at all. This is OK here. A gay bar is different.

F: OK. Pretend that the bookcase at the end of the room is the bar – some other group members please stand by it, pretending to drink. As you **F**, stand and watch, what do you remember of approaching the bar and the people standing there?

F: That's not my memory at all. I'm remembering my first day at school – my mother left me with all those terrifying people.

E talked for some time, bringing his early experience into the present and at **F**'s suggestion re-evaluated it in terms of his adult self and his present abilities. The following week **F** reported entering the Colherne and finding it very boring. Nobody spoke to him and nobody pursued him into the toilet.

And lots more – each a pair of polarities. And, as both L. and F. Perls emphasized, the affected person needs to know both the pain and the self-generated balm.

Meanwhile, the ideas of Wertheimer, Koffka and Köhler, summarized in Table 1·2, lead us to expect wholeness which is absent if the experimenter is not at all curious about anything. We can also expect a lively, consistent cross relationship among activities, so that if the person is preparing a meal interruption will always be followed by a return to cooking activities.

Is this person exploiting his situation to his maximum advantage, using thoughts and feelings congruently and beneficially? Is this person doing what he/she has always done or are there are signs of an experimental attitude and risk taking? If I suggest an experimental change to what is happening, how does this person react – with hostility, with interest but no action or with interest and a change of behaviour? Is this person aware of his main interest and centre or does he direct himself onto peripheral, inconsequential avoidance activities?

To explore these aspects of behaviour I will now consider deleterious events at the figure stages, using as far as possible, the criteria of Table 1·2 and thus of Wertheimer, Koffka and Köhler. And also resume discussion of interruptions begun on page 56.

A group has been very active and quietens down ready for the next volunteer. His story is summarized in the box called **First Fool** which contains a story that is very typical of

Somebody says he/she needs help. What is she/he likely to be like? What sort of character and behaviour will the facilitator see?

This is an odd person indeed. So locked in deleterious, self-defeating behaviours that one needs to remember that he/she is barely aware of what is going on.

- Instead of being terrified by the prospect of meeting powerful people he has the security of shyness and boredom.
- Instead of pangs of hunger he has the addictive responses of the tobacco smoker.
- Instead of facing the wrath of a wife he has alcohol inebriation: "pity me!"
- Instead of spending time and money on a wholesome meal he has the pap of instant junk food.
- Instead of a country formed fresh air complexion she smarms herself with cosmetics to falsify her appearance.
- Instead of going with one interest, attention wanders randomly and chaotically or concentrates obsessively.

Chart 3·2. A double embedment

The first figure is shown with an interruption at stage <3> and the second at stage
<4>. Reaching a satisfactory consumma-
tion of each figure is necessary before
return to the previous figure

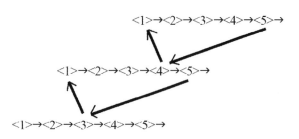

---

presentations by suburban young people. **E** had a future
projection about being seen as a fool – enough to make anyone anxious. And here was the test bed – would **F** and the
group members see him being foolish? No one did. This, in
terms of figure stages had him mostly at <2> with ventures
to <3< where he felt tense in his rib cage. He found he could
relieve the tenseness by breathing and laughing and so reached
stage <5> where he questioned group members.

In the story called **Problems with men E** said she was
anxious and was clearly worried about men who she knew
were all like her husband. **F** saw the process as being reality
testing. While whining she was stuck at stage <2>. Stage <3>
occurred as she remembered times when she was OK and
then situations when men were OK with other women. The
active stage <3> exploration of options came as she ran a
dialogue with a man who was evidently considered to be OK
by another woman. And she did not feel anxious about him.
Her main figure was about men and it hovered around stage
<3> until she discovered Boss.

The story untitled **Lithium** ran for a long time. Here **E1**
was living a fantasy and his interaction with group members
provided the way for him to find reality. He only talked so
he was probably stuck at stage <2>.

Fear of the future is most frequently dread of something
happening. In the story entitled **Toilet training E** was like
this and was very agitated as he slowly revealed his dread of
going into a gay bar. He was about 20 and had a strong
impulse to do so. Here **E** started with an apparent fear of a
future event. **F** encouraged him to behave as if he had done so
satisfactorily, stages <6> and <7>, and "remember" what
had just happened. Instead **E** remembered his first terrifying
day at school, another figure, loaded with anxiety, at
stage <2> and <3>. This was enough for **E** to detach himself
from the memory. The memory was only pungent if not in
awareness.

E in the story entitled **Father problems** had been saying
in group how miserable she was. On one occasion she got up
while talking and walked around until she reached the fire
place on which she rested her elbow. The dialogue proceeded
as related in the box.

Over some further weeks in group she shared instances
when she had been miserable and suddenly remembered her
father whereupon she changed her feeling, usually not
towards happiness but to anger at her father over what had
happened and sometimes to sadness about a relationship
that could have been warm within the family. Her
process of self-nourishment was one of freeing herself
from automatic entanglements with her father that had started
in childhood when she had little choice for herself.

Going in reverse

E in the story entitled **Dummy** is in a figure at stage <5>
presenting a spoiled impression of himself – a message from
the past has bothered him and he doesn't know what the
message means. Something has concerned him about events
in a supermarket but **F** regards events in the group as being
more relevant to the development of a complete gestalt so he
decides to encourage **E** to go into reverse on the matter which
is fresh in **E**'s experience. The starting point was **E**'s particular now of that stage <5>, messed up action

From there a slow progression backwards through
engagement, (with his father) to the initial excitement spoiled
by expectations of being thought stupid, stage <2> where he

Dummy

Ewan: I thought I would tell you about my trouble in the supermarket but instead I am feeling shy and embarrassed.

F: You can explore those feelings, shy and embarrassment if you want to. Tell me in as much detail as you can carefully manage. [silent pause]. What is going on in your body?

E: Trembling in my lower belly [puts his hands over his diaphragm]. I feel a bit sick in my stomach.

F: Do you want to vomit?

E: Oh no! Not as bad as that, just a sort of revulsion feeling.

F: What don't you want?

E: I want to be free of these bad feelings.

F: Would you like to experiment with your feelings?

E: OK. What do you suggest?

F: As you stay with your feelings of revulsion, tendency to vomit, perhaps, shyness and embarrassment, you can also become aware of what happened for you, immediately before these discomforts. Just take your time.

E: [after the minute or so] I had to push myself to speak up and take a turn in the group. I was aware that somebody else might get in before me and felt sort of generally excited, though reticent. The trouble in the supermarket was yesterday …

F: I notice that instead of staying with your present discomfort, you have escaped back to yesterday. I suggest you stay with what occurred for you as you struggled with yourself to go public in this group.

E: Yes, OK., I feel clearer now. I am speaking out without pushing myself.

F: Look around at us. Say something to some of us.

E: [To **X**] As I was struggling to get started I thought you would think me rather stupid.

F: Turn that statement into a question.

E: Would you have …

F: … in the present tense.

E: Do you think I'm stupid?

X: No. Quite the contrary. I enjoy what you say and I'm with you as you share your feelings.

E: Oh thanks. I wonder why …

F: How are you doing with being in touch with events as you struggle to present your problem to us.

E: I was rather thinking about something else. What **X** was saying. For me, I'm still bothered by what other people say about me. My father called me Dummy when I got something wrong and he went on and on about it, as if I would eventually get things right. And I remember like, I can hear his voice Dummy, Dummy, Dummy.

F: Can you say that to him, now? Imagine he is here with us - Dummy …

E: You shit. You were the Dummy. I was a only little.

E then got into a cushion bashing session while shouting to his father, Dummy, etc. After a few minutes –

F: **E**, what have you to say to yourself?

E: I'm not a Dummy.

F: What are you?

E: I'm OK. I know you like opposites so I'm a wizard, a wisey, I who sometimes makes mistakes.

F: Tell us, each, here, what you know about yourself.

E: [To **X**] I'm perfectly capable of getting things right. OK, I make mistakes sometimes like any one else but if I am left alone I'm OK.

F: Who should leave you alone?

E: My father and other interfering adults, like teachers, they were just as bad.

F: So in the future, as you want to assert yourself, what will you say if your Father's voice says Dummy, [sing-song] Dummy, Dummy

E: Fuck off, I'm OK. I go my own way. [he looks fiercely around the group].

F: You are trying to frighten me. Do you need to?

E: [calming down] No, it's him, my Dad I need to frighten off, not you, not you lot. [laughs deeply while looking around at the group members].

---

remembered his fathers "Dummy" injunction. With this in mind he dealt with it and asserted and associated with his basic OKness. At this time also he exerted his reaction to the injunction, or so **F** thought, as he drew attention to **E**'s fierceness, which like Dummyness, was at first projected on the group. This is the dialectic at work, a jump from passivity to aggression.

Finally **F**, who was feeling pleased with this general outcome, summarised the state of the current figure – in the group – and possible future figures, since **E** has learned to protect himself from his Father's injunction and needs to have this state of affairs reinforced so that it will always

operate in the future. Or rather, so that the injunction becomes inoperable. And yet there may be times when he will like to be a Dummy as when playing with his children.

Interruptions and embedded figures

An interruption of a developing gestalt figure is a point of formation of a new gestalt figure. The first figure is not completed unless one decides to go back and complete it. Multiple interruptions can result in chains of figures and a muddled situation as illustrated in Chart 3.2.

From the facilitators point of view the way out of the muddle would be to systematically go through the figures in reversed order. This would be analogous to the Lankton's procedure[11] for dealing with embedded metaphors. I have dealt with this situation in therapy situation but with no clear conclusions as to effectiveness. My problem was remembering all that had gone before so that I could effectively guide the experimenter back along his interrupted experiences. Facilitators with younger memories than mine may well find happiness when dealing with interruptions by noticing embedments.

Is there a future for gestalt figure stage analysis?

My experience indicates that figure stage analysis is a valuable asset, is a smarter way of facilitating experimenters processes than looking for introjects, projections, etc. But these belong in the next chapter – so here goes!

Notes and references

1   Enright, 1970b, pp. 263 → 273.
2   Stevens, 1976.
3   Paige, 1979.
4   Perls, F. 1979. And see the Appendix, page 172.
5   A routine procedure for developing awareness is set out on page 172.
6   Goulding & Goulding, 1979.
7   Use of the figure stages to help a phobic person is set out on page 172.
8   Perls, F. 1977, p.80.
9   Coué, 1926, whose affirmations develop awareness as well as priming protofigures.
10 Polster and Polster, 1973, p. 179.
11 Lankton and Lankton, 1983.

The development protofigure
follows the same track as set out here
as will be explained in Chapter 9.

—=(☆)=—

Chapter 4
Secondary Failures of Interpersonal Engagement

A module of neuronal tissue
becomes expressed and recognized in awareness
as an aspect of a gestalt.
Such a module is known
in the ground of the gestalt
as the protofigure.

Two people engage in conversation and/or other activity with ease or with difficulty as when one person breaks engagement and so makes the other person feel uncomfortable or actually hostile. S. and A. Freud[1, 2] in 1937 and F. Perls[3] in the 1950s categorized these styles of failure of engagement naming the important ones as introjection, projection, retroflection and confluence. E. and M. Polster[4] in 1973 added another, deflection. Yet another, egotism, is sometimes considered. These were described as defence mechanisms or resistance processes and will be examined in detail in this chapter after recognizing that G. Wheeler[5] in 1991 showed that these apparently deleterious contact functions, as he called them, had beneficial aspects and should not be considered to be entirely negative in effect.

Association

I prefer to consider these activities of making or breaking interpersonal relationships as engagement styles and to categorize them in relationship to the neurological activities of association and dissociation (pages 6 and 54).

Beneficial neurological association implies intact innervation of the various modules involved so that they are neatly interconnected, facilitating complete functioning. Psychologically each function is an action towards satisfying needs and/or the side issues of appetites, curiosities and wants.

For the happy person, if his or her association engagement styles are personally satisfactory, he or she feels wholesome and has ideas about logical completeness, as a gestalt would be. For myself, if I'm not totally satisfied intellectually with what I am creating I can fallback and say that I am creating a metaphor*.

This chapter is mainly concerned with deleterious process and how to attain a state of happy venturing. However

most of life is concerned with joyous adventures, as stressed above, and the relationship between engaging peoples can be described, colloquially, as "good chemistry".

Dissociation and alienation

However, this chapter is devoted to habitually occurring situations that are not beneficial, is devoted, in Freud/Perls nomenclature, to interruption processes. When dissociation occurs and a gestalt process is interrupted and not completed, a stuck state and/or an alternative unwanted state occurs. This is certainly characteristic of trauma states [it may not be clearly delineated for other states]. The person sets out to do something or other and that process is interrupted and aspects of the trauma, perhaps a flashback, supervenes and the person is unable to do what was originally intended. A more frequently occurring example concerns the person who repeatedly fails examinations. He or she intends to be cool, calm and collected when sitting at the exam desk but actually becomes excited, panicky and distressed. An event or events in the past repeatedly interferes with present behaviour.

The person who is always late when meeting other people gains a reputation based on that fact because everybody else is fed up with being kept waiting. We can do with a sub-category of dissociation like *tardy* to label this person and, relative to this person, it would be as important as introjection, projection, etc., but not be subsumed under any one of those concepts.

There are many other such terms that could be used for designation but it is convenient to deal with the historical ones first and then consider a few more that don't fit the pattern set by the Freuds.

The key concept is the same, however. A personal habit is worthy of note if it disrupts association, alienates other people and causes the other person to dissociate and put his or her attention elsewhere.

---

* Metaphors have been shown to be neurologically homogenous with the events(s) they relate to. Chapter 10.

When I am working as a facilitator I have two centres of awareness; what is going on for me in this relationship and what, do I surmise, is going on for the experimenter. Then I am interested and ask -

> 'What do you intend to have
> for yourself, here, now.'

From that point on the experimenter is in control of what happens though I am concerned to facilitate the experimenters process by using clean language (page 163) and I watch out for ways that he or she disrupts our engagement.

The out of awareness module of nervous tissue that is precursor of the gestalt figure (ground, a protofigure, see Chapter 9) becomes chosen by some unknown selection process and is expressed as an engagement style. An explorer in the jungle meets a tiger and, without thinking, reaches for his rifle. The selection process is automatic, spontaneous and life preserving. The engagement style is the result of previous experience – the explorer went to a special school to learn how to use his rifle and track game.

In another example is the person who, in a situation with many people being organized to combat peril, tends to panic. He may say to himelf: "Don't panic!" and also tends to energetically tell everyone else: "Don't panic!" His first personal reaction was to become excited in a disorganized way. His second reaction was to project his excitement on the other people. He will repeat this behaviour whenever he finds himself in that kind of situation and his colleagues will know him as doing just that as he or she also confuses them by using negatives (page 178).

Satisfying needs by venturing out in the environment is essential for the continuation of life by bringing in needed things by *introjection*. Sometimes incoming material is not wanted and is *deflected* away. The venture out is called *projection*. Alternatively a personal possession is not *projected* out into the environment, it is withheld in an act of *retroflection*. Also to be considered are *confluent* acts that occur when the person is unclear about his or her separation and difference from other people. *Egotism* is said to be a state of complete domination by a person but I have not come across it.

While the following descriptions are mostly concerned with single engagement styles it is clear that several may exist at the same time or in sequence, and certainly many form anti-pathetic pairs. There are always at least two

people involved when engagement occurs. To emphasize again, both may benefit during the encounter or one or both may suffer.

To emphasize, we are considering here deleterious procedures that occur as commonly as do valuable procedures such as saying "Hello!" and extending a hand for a shake on meeting or sharing a "high five." Each behaviour is regarded as habitual and characteristics of the person, so that, for example, the man who loses his temper when dealing with children rarely does anything else when meeting children. He will have excuses (mythology) that he present to himself and anyone who will listen.

Describing a person's deleterious behaviours may become a presentation of stereotypes: "He's always so aggressive!" "She's so passive we hardly know she is here."

This chapter is devoted to dissociation and synonyms will be used in the style of alienation, interruption, disruption, disconnection, disengagement, block and failed association.

Successful and unsuccessful ways for finding happiness

Personal problems occur and further problems result from attempts at correcting unwanted behaviours (see the Anna Freud box) – not everyone is a successful self-therapist and misery is perpetuated! All people seek happiness and avoid anxiety if they can.

Set out hypothetically: a particular engagement style may work well within the family of origin but when the person begins to work among other people he or she may become rebuffed and hurt. Another engagement style could be tried and may or may not be satisfactory in the new environment.

If the hurt results in the person feeling vindictive, the new engagement style may have negative consequences, arousing antagonism among colleagues. However our person may feel "one up" and successful in getting what he or she wanted and the second engagement style may become habitual. Ameliorative action has then to deal with both secondary and primary faulty engagement styles and provide an effective style for future use.

A common example of this process concerns a depressed person who tries to solve his or her problem using alcohol. Depression was the first engagement style and inebriation the second – neither style fits in the categories discussed here, which illustrates the need for recognition of a much larger gamut of engagement styles.

Notes on the traditional engagement styles

The sovereign principle
which governs the psychic processes
is that of obtaining pleasure.

Anna Freud, 1937, p 7

Introjection

Beneficial introjection. The prosperity of a species depends on intake of nutrition including water. There is also intake of oxygen for haemoglobin to carry to the tissues and of ideas about the whereabouts of food, water and

fun. Each item taken in is subject of metabolism. Food is digested down to basic elements, amino acids, fatty acids and glucose before being catabolised to form the organism itself.

Ideas must also be digested. "The deer are on the river bank" – which river, which bank, how far away, who is coming too, is there also a tiger hunting there?

Introjection is about taking things in, receiving, and is the most prominent engagement style because, as stated above, nutrition depends on it as does the gathering in of ideas and also finding a mate. Both processes are biphasic in health – food is a useless lump in the stomach unless digested and assimilated. Ideas likewise must be mulled over and made the ideas of the person involved, a process rather like immunological imprinting. Students use this process with note taking in the lecture theatre followed by consolidated at home as the notes are read, thought about, remembered and written up in the style of the student. Students accept the teachers projections.

Having found what I need out in the environment I bring it back into my vicinity and do what I need to do to satisfy my basic need and that might be some kind of preparation or cooking.

Other people may have been involved in that activity because out in the environment I sometimes find what I need in the hands of someone else. No longer having to dig in the earth to obtain vegetables or hunting to obtain meat I have to barter or pay money for the vegetables and meat.

Deleterious Introjection. In the extreme case the person loses the ability to evaluate what he is doing as he brings home things he doesn't need. I remember a man who retained every newspaper he bought: "I might need to look something up" was his excuse and myth. A group member reminded him of the fire hazzard attendant on storing such a lot of paper. I had him talk: "I am your stock of paper and I want you to know something." "What do you want me to know? I am going to choke you to death."

I take in what I want plus stuff that "leaks" in when I am off guard. These are intrusions which get in my way unless I attend to them and really make them part of me.

Politicians set out to intrude, especially at election time, subverting peoples views although the lively citizen resists.

Receiving and being imposed on. Most people are pleased enough to receive a present provided that there is nothing nasty attached. A teenager who is given a car for her birthday will certainly be miffed if it is conditional on her not seeing her current boy friend.

Engagement becomes awkward following introjection, is felt to be insincere and not genuine by both parties. "The maturation of human beings necessarily involves the challenging and assimilation of introjected values thus asserting one's powers of discrimination and stand-

ing on one's own moral feet." (Swanson[7]). Perls[8] refers to this as moral regulation and notes that muscular tension and lack of sensitivity and feelings result.

The alien stuff of an introject is not associated with the self. It is associated with anxiety and any amelioration occurs by digestion and assimilation thus associating with the personalized product.

Introjection impedes growth because growth always requires *assimilation* from the environment, as emphasized above. Using food ingestion as the metaphor again, what is taken in, unchewed, not digested and not vomited hangs around as an uncomfortable lump. Perls, *et al.*[9] described assimilation as aggressive – it is not. Digestion of food is assertive. Aggression all too readily turns an introjection into an anti-introjection that can be just as harmful.

When I was a boy I had to go to sunday school and learned a lot about Jesus and his friends and enemies. I was never convinced of the validity of the stories. I had in memory a lump of information that did not belong to me and, in fact, really still belonged to the teller of story. The lump of data was an unwanted introjection that, in my 'teen years, I rejected.

My father told me that it was OK to steal by saying the old soldiers phrase: "Anything not tied down is mine." I knew he stole things when he was in other people's houses selling insurance and for years I had frequent ideas about stealing things that were lying around. After 70 years I can confess that I, as a teenager, was often tempted to emulate my father's actions but very rarely succumbed.

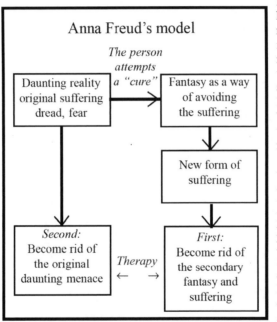

Anna Freud's model

The person attempts a "cure"

Daunting reality original suffering dread, fear

Fantasy as a way of avoiding the suffering

New form of suffering

Second: Become rid of the original daunting menace

Therapy

First: Become rid of the secondary fantasy and suffering

Anti-introjections. An anti-introjection is exemplified by the obdurate son who refuses to do anything his father suggests although he knows that some of the things would be beneficial for him.

If "be quiet" was frequently shouted by father it is remembered and is an introjection. If the child reacts vigorously in a temper he or she may create an anti-introjection. "I won't, won't, won't be quiet!" The anti-introjection is as effective as a disrupter of behaviour as was the introjection that originated it. These, introjection and anti-introjection, are polar opposites (page 28) like F. Perls' top dog and under dog and can be resolved by two-chair enactment.

Rules of behaviour. Another way of looking at introjections is to consider rules. Whoever plays a game learns and accepts the rules. If he become a prisoner he has to accept the prison rules or be put in solitary confinement. These rules are introjections from social systems.

Conscience is the name given by some to the mass of introjections provided by the social group of one's origin. F. Perls[10] refers to the work of self-manipulation required to deal with conscience because consciences always remains as if they are internal and they remain "internalized external controls." They are also authoritarian because "consciences behave as if they were an absolute morality" (F. Perls[11]). The engagement boundary in introjection is well inside the physical boundary of the body and being external, conscience can receive other people's projections such as anger. It is usual for moralizers to refer vigorously to the wrath of God. So conscience is always aggressive says F. Perls[12].

Dealing with introjections. F. Perls suggests that the facilitator can lead an experimenter to clarify his experience by "interviewing the introject" as a two chair dialogue between self and the intruder. The imperative words, should, must, etc., can be changed, in experiment, to the assertive word, "choose" to determine the effect of the change. Imperative forms of verbs are not used. Intrusion is the primary process of learning anything, and the balance of the person in the experiment is only re-achieved if a process of digestion occurs immediately or later: "is this really mine – what do I have to do to make it mine?"

Deflection

Beneficial deflection. My personal interest may depend on me being not interested in dealing with what you say. I talk about something else although I may intend to get round to your query in due course.

The deflector has a habit in which he or she takes little notice of the other person's presentation and goes off on a line of his or her own, dissociating from the "intruder" and feeling right about doing that. The presenter is frustrated and probably annoyed that his or her proposal was not acknowledged. The deflector may have had the sense that he or she was working round to what the presenter wanted, but that may not have been evident to the presenter. One sees politicians doing this on the TV when the questioner puts an awkward question. Of course, the proposal might have been rubbish for the deflector but he or she might have been polite and said so.

Deflection, as described by Polster and Polster[13] is as the term infers, a rejection of incoming material, usually without acknowledgement. This has aspects of alienation because the presenter wonders what has happened. It is not in itself an interruption or avoidance process because the receiver goes on with what was happening anyway. Spontaneous deflection, likewise, saves the receiver of attention in awkward moments and is O.K. if immediately followed a disclaimer: "Hey, I am not interested in that. I want to talk about . . ." Otherwise the relationship is spoiled.

Deleterious deflection. The person who is rebuffed by a deflector may feel hurt. If you ask me: "How are you today?" and I reply "Old people always have some kind of ailment." I have completely discounted your interest in my health, etc., and gone off on my own track. A polite reply would have been something like: "I would rather not talk about myself, just now. Is that OK with

---

Drear

On beginning to engage with the facilitator, the experimenter presents in an engagement style that was probably familiar and habitual. He started talking with a dreary account of how nothing went right for him. Only later, as he became more excited, did he complain of lack of support from the facilitator, a simple projection as he was actually failing to support himself. The facilitator had best give attention to the engagement breaking, failing dreariness before the projection.

---

you?" and it might not be or be followed by discussion. Questions are often used in a lazy way to deflect, especially to get someone else to do the action. Talking about rather than to the person is diffuse. Generalizations and diffuseness are ways of avoiding excitement and responsibility.

The deflective reply is often a generalisation as in the "Old people" example above. And: "How is your mother?" "Mothers are always such a bother." Deflection is avoidance of emotional engagement by some kind of irrelevant activity thus dissociating and discounting the interests of the other person. However, the deflector will have gained time for thought.

Circumlocutory talk may be deflective, but deflection is not always verbal; it may be a temporizing action like lighting a cigarette. Negative statements are often deflective in character as they present the polar opposites of the positive aspect and what is negated.

Body language in deflection is often very clear, the deflector may turn away and loses eye contact with the donor. Voice tone may become monotonous. Gestures may change from intimacy to encompassing the horizon.

Dealing with deflection. Polster and Polster[14] say that this condition is best dealt with by encouraging direct engagement. They asked a deflecting person "to make up several statements beginning with 'you'."

When I was becoming aware of my deflection processes and the effect they had on other people I set up an experiment for myself in training group. I decided to become aware on the moment when I deflected while also being aware that attention to deflection was itself a deflection. After two days I stopped the experiment and continued to be myself. The feed back given me was that I was making much clearer engagement with fellow group members This procedure is recommended by the Polsters who also recommend talking about what was actually happening while deflecting. They recommend devising experiments in which deflection is deliberately used, thus "getting the feel" of it.

A facilitator who points out something while disregarding the existential state of the experimenter is deflecting, e.g., he remarks about body language while the experimenter is talking. It is then the responsibility of the facilitator to return the experimenter to his interest: "While you are talking you can also be aware of what your feet are doing."

An exercise for a deflector is to have him or her going round group members, asking them to challenge him or her. Then he or she replies. Deflection, if it occurs can be corrected immediately by associating with the other person instead of dissociating.

Projection

Clark, (1989, p 103) traces the origin of the projection idea to scattered remarks by S. Freud that were consolidated by A, Freud (1937). Clark saw projection in a wide context because whenever people engage with one another they make guesses about what is going on for the other person and act on the guesses. The actions are projections and all goes well unless a mistake in interpretation has occurred. Such mistake include events when the projector has assumed that his/her oppo' is the same as him/herself. This is "... attributing intolerable behaviour to others that is characteristic of oneself." Projection is an act of dissociation.

A person has characteristic, stereotypical ways of projecting which arouses antagonism in the recipient.

Heard in Bridport High Street: A boy of about 5 or 6 says to another boy: "Arse 'Ole!" The other boy replies: "Stop talking about yourself!"

Beneficial projection. The prosperity of a species depends on ejection of the waste products of digestion, as faeces from the anus, in water from the kidneys, as CO_2 from the lungs, etc., and in aqueous solution from sweat glands.

Shitty ideas can be rejected by disinterest or by "sharing" with another person. "I am angry with you but I see you as being angry with me." A typical projection.

A person stuck with extruding verbal abuse, for example, may wonder why people avoid him and he has no friends – but he may not care.

The end of a quarrel is ejection of the opponent. At the end of the game the players disperse, sprayed in the various directions of their homes.

Deleterious projection. Projections are often introjected material reappearing. Thus if a man is among misogenyst men, he may, carelessly for sociability reasons, join the exchange of anti-women yarns and taunts.

The receiver of a projection has gained a replicate of rubbish the projecter has rejected. If the receiving person is fully aware of what is going on he or she will reject the rubbish.

Racists are typical projectors. They have fantasies about other people and then say: "I told you so" when the other people defend themselves. These are dangerous ejections and prejudices, anti-Semitism, anti-Catholic, anti-Protestant, anti-Muslim, misogyny, anti-Gay and anti-Het, all may lead to death and destruction.

Projection is an ongoing process – that which is within is duplicated outside and there is no limit to the number of replicates. It is quite usual for a father repeatedly to try to force his son to be a replica of himself.

Perls, *et al.*[15] considered that projection happens out of awareness – if I dislike **X** I experience him as disliking me. Projection kills excitement because no energy is sent out with the idea. However the message is received and may induce an excited reaction.

Evil is frequently seen in other people by projectors of personal error. But "as Socrates said long ago, evil is simply error."[16] As I say to myself "If I am adventurous I am bound to make some mistakes and I am O.K. with that, having made restitution if that was necessary and appropriate."

Gabby, during supper, had given **Joel** an uninterrupted lecture lasting about 10 minutes. **Joel** protested that he had been unable to get a word in edgeways. **Gabby** said: "You never listen to me!" to which **Joel** replied angrily: "I have been listening to every word you said and very interesting it was but I wanted to ask some questions on matters that I did not understand." It was, of course, **Gabby** who had been talking and not listening – he replicated his condition on **Joel**, much to **Joel**'s consternation.

In Prokofiev's opera "The Fiery Angel" the principal singer spends all her time searching for said "Fiery Angel." It is clear to all the other characters and the audience that she is herself the Fiery Angel.

Every time I engage with anyone I am presenting myself for scrutiny, timorously or boldly. Pathological projection is something extra as I put aspects of myself on the other person, expecting him or her to be as I imagine. There is an imperative or a "should," here. The pathology comes about because the other person is not what, rarely who, I expect – if he or she succumbs to my projection I am into power play, being aggressive. If he or she does not succumb I have lost my power play and I passively retreat hurt and disappointed.

Perls emphasized the moralistic and judgmental aspects of projection, "of disowning aspects of oneself and ascribing them to the environment, making the environment responsible for what originates in the self." (Swanson[17]). Environment is partly other people. And "the resulting evaluation takes on a rigid and absolute quality." I add authoritarian and aggressive quality. Whereas the introjector takes oughts, etc., from other people, the projector throws them on other people. Replicative projection can thus be a form of self-protection.

The Freud/Perls term projection derives from an inadequate metaphor because a projectile once gone is gone; the shell case is empty, the projectile exploded, both unusable and not suitable as metaphors. The next projectile is not the initial projectile. To speak of missile projection is to infer that that game is over and won't be repeated whereas the contrary is true in human relationships and whatever the psychological explanation is, can be repeated an infinite number of times.

Dealing with projections. Everybody extrudes and projects and then rationalizes about his or her actions[18] and, perhaps the most important aspect is that it, like a lie, changes the environment, changes the context. This may induce the experimenter to become afraid of the vengeance from out there – is it a coincidence that current social mythology is so much to do with aliens from outer space?

Perls set out steps for dealing with his projections[19] which I modify and set out as –

• Realize that the projection exists

- Recognize that the projection belongs to me
- Assimilate the projection

The latter is particularly important because introjection of the projection leads to paranoia. Paranoia is attempted self defence against incursions which, are extruded again, this time against any, though irrelevant, target that is to hand. These bursts of temper are projections of shame, fear, guilt, grudges and embarrassment with moralization and victimization.

Assimilation involves –

- Recognition that under some limited circumstances the projection is true. Act out using two chairs. In terms of a bogey – be the bogey; the original engagement styles (resistances) reappear and can be owned. The pacifist can come to own the sadistic, militaristic streak in him
- Become aware of the mythology, ideas and scotomizations that arise to justify the extrusion.
- Emphasize the reversal: "The suspicious man should suspect himself," the victim victimizes others.
- Ideas about jealousy, suspicion, injustice, victimization and timidity indicate projecting.[20] Guilt expresses regrets about the power of the introjection. I hurt you by throwing that at you – sorry!
- Acknowledge needs for love, and to express hate, resentments, etc. – and other people can love me and hate me.

In all, the dissociations of projection are changed to associations.

Self-consciousness has many names. The person is said to be shy, timid, bashful, embarrassed, reserved and/or reticent. Whatever the euphemism the person is avoiding engagement with other people The therapy of self-consciousness involves fantasy in which use of all sensory systems lead to the defeat of the object person[21]. The power of expression returns, and with it "confidence, a new ability to deal with difficult situations, and recognition of … the environment." "One way to cure self-consciousness is to change it into object consciousness."[22] Such self-consciousness is certainly not an "inferiority complex" at work.

The curative steps for self-consciousness "are obvious. You have not only to become fully aware of what emotion, interest or image you are concealing, but you must also express it in words, art or action."[23] Self-consciousness arises in particular dangerous situations, and is often "cured" by drugs, alcohol, nicotine, etc. Instead "change the wish to be admired, the fear of being stared at and the feeling of being the centre of interest into activities of being enthusiastic, of observing and of concentrating one's interest on to an object."

If the experimenter says "they don't like me" suggest turning it round to be "I don't like them." "They're always trying to push me around" becomes "I want to push them around[24]".

Perls[25] cites an example of an experiment –
Perls: 'Can you imagine what my response would be?'
Client: "What a horrible creature you are!"
Perls: 'Can you say to me – "what a horrible creature I am'?"

A good neighbour

A lass came to see the facilitator after suffering a "nervous breakdown." Her history included an account of an absent father, perpetually ill mother and four younger siblings to look after. Her 24 hour devotion to the family was interrupted when a neighbour came in to help. "She was taking away my children!" The lass felt that all her inadequacies were reinforced in the presence of the highly experienced neighbour and she stopped being devoted.

Facilitation started with reinforcement of appreciation of the benefits she gave the children and her mother and continued with an investigation of how she might give more time to looking after herself. She was also interested when the facilitator pointed out that the arrival of the neighbour brought behaviours she might emulate instead of seeing as threatening.

Client: 'Yes that is what I intended.'

When a grudge is expressed, visualize the person towards whom the grudge is directed[26] and be free in expressing the resentment together with feelings. When calmness has returned, face the fact that the adversary was not the fantasy person but was in fact oneself.

Dreams.

In actual practice I let the experimenter act out all the details of the dream . . . We assume each part of the dream is a projection. Each fragment of the dream . . . is a portion of the experimenters *alienated* self. (F. Perls[27])

F. Perls wrote that dreams are expressions of alienated, projected fragments of self that need to be recovered. That seems to me to be an idealistic, non-concrete expression of events. The dream may be excreted but the memory remains.

Dreaming is a gestalt completion process. It is only when completion does not occur, as with recurring dreams, that therapeutic intervention may be necessary. And then with the intention of attaining completion. Dreams are much more than projections, they are largely contorted memories. They are also shit and are best excreted and finished with, even if therapy is required to diminish exaggeration and attain completion. Only recurring, obviously unfinished dreams are worthy of the attention that Perls[28] gives them. I personally don't divine over turds. Extrusion is excretion: do I shit in the toilet pan or on you – or on myself!

My attitude when someone wants to "work" on a dream is to suspect that something else is being avoided and to say: "You are interested in your dream. Find out what you can make of it, be aware of your feelings as you go along and ask me to make suggestions if you want them".

Otherwise dreams well up from ground states and are effectively distorted figures. I dreamed that I was at a table with many men who were eating but I had no food in front of me. I woke to find myself feeling very hungry. Such a dream needs little analysis.

A dream is a fragment of a figure, no matter how distorted, so it may be possible to reach completion by reconstructing the rest of the figure. If an experimenter focuses on activity, the facilitator can be interested in "what happened before that" or "what happened next." The trancy question[56] "I wonder if you are interested to investigate what happened before or what happened after your present experience" can be very useful.

In the last paragraph I showed my personal interest in action as a likely centre of a dream. Any other stage in the figure circuit (frontispiece) is available and turns up. The following are common; engagement without satisfying action accompanied by waking with vague feelings of disquiet and fear, awakening inaction with feelings of unease and vague memories of some previous disturbing event.

Dream work. When acting out a dream get into the posture of the dream: usually lay down covered by a light blanket. After listening to the dream the facilitator can say: "So all that happened to you!" and either, as said above. "What happened next" or " What would you now do instead?"[29]

Igor said that he had an interesting dream but doesn't remember the details of it. The following is based on F. Perls method somewhat updated. The facilitator encourages the experimenter to say: "Dream, I will not remember you" or "I will not under any circumstances remember you, dream." *"Now reply as the dream."* "I am **Igor**'s dream and know that I was eating a banana – it doesn't taste at all and I won't chew so it can be swallowed." *"Be the banana."* "You are not going to mash me up!" This was a penile fantasy dream. He wanted his girl friend to suck his cock – like his boy friend did – but was afraid to ask her.

The sad thing about projection is that the projected material is not gone for ever because it was only a replica that was extruded. A residue remains to bother the person concerned and be extruded again, repeated as often as suitable.

Retroflection

Beneficial retroflection. My personal interest may depend on me being about to project something on you but I withhold it. I may suffer what I expected you to suffer but more likely I change the style of suffering and feel ashamed that I avoided engagement with you. I may tell myself that my ill-feeling is guilt.

How does the perpetualy self-defeating person benefit by bringing back on self the shame, blame, etc., he or she might have put on someone else? At first he or she has avoided the risk and danger of engagement with this other person and may feel pleased. Then the realization that he or she has disrupted the relationship with the other person occurs and the soreness sets in: "I'm ashamed of my cowardice. I blame myself for timidity." Retroflection is an act of dissociation.

The potential receiver of the action may have had a vague sense that something nearly happened but is otherwise unscathed. Here the excitement and affect are engaged out in the environment and fear of consequences

induces an interruption – a pull back into self occurs on the only safe object in the field, him- or herself. Fantasy builds up regrets over the failure and the energy that was not expressed locks into his musculature producing psychosomatic effects, head aches, etc.

F. Perls, *et al.*,[30] point out that intellectualization is OK in the generation of beneficial retroflections, as he called them. For example, stopping the impulse to go into the sea is life saving if a storm is brewing. Self-defeating activities can be valuable temporizing processes but become deleterious when associated with persistent muscular tension because a tense muscle eventually aches with cramp.

It is helpful for a facilitator working with an experimenter to investigate muscle tension by suggesting deliberately tensing a muscle or group of muscles. There are times when relaxation is effective and alternating tensing and relaxing is even more effective. While this is going on ideas may come up? Here is a clue list –

A clenched fist can indicate	I want to hit you
A cramped neck can indicate	I am stubborn
A stretched neck can indicate	I am haughty
A catch in my throat can indicate	I want to cry
A tight throat can indicate	I won't swallow it

Surrogate activities are a great help with these and other indications. Instead of hitting **X** the person hits a cushion while talking as if it were **X**. Otherwise the cushion can be strangled, stretched, squashed, beaten, as is appropriate for what is to be discovered while expressing affect and knowing clearly that "It is I who am doing this!" During this process the appropriate memories from the pas t arise spontaneously.[31]

The Polsters[32] emphasise that "thinking is a retroflective process which is only acceptable if producing need satisfying action." Thus self-defeating activity is only satisfactory if beneficial projects are replicated outwards, refused and then brought back on self by the interested person. "You refuse my love and I want you to know that I love myself" – a usual reaction is to hate self for failing. The latter is self-punishment. Self-defeat occurs as I test myself, as I venture out into my environment. I make mistakes, I retreat and recognize the self-defeat.

Detrimental retroflection. Perls, *et al.*[33] also express concern about the split that occurs on questioning self: "I ask myself ...." I regard this as beneficial for me since it is only a process of clarifying thoughts and not necessarily a matter of avoiding questioning someone else. It is a form of self-responsibility. It is also akin to role playing parts of self by speaking from cushions an exploration of options.

Self-defeat by retroflection is simply the state in which I shit on myself instead of on somebody else: If I want to make the comment: "You should be ashamed of yourself; you've not washed your face in weeks" so I am likely to say to myself: "I should be ashamed of myself for not speaking out!" This phenomenon is common, disruptive of self-nourishment and disruptive of engagement with the other person who does not know what is going on and yet is sensitive to something happening. The process was called, by F. Perls, retroflection as if the initial statement had been projected out and then drawn back again: retro- means bringing back. In fact, no projection and no movement occurred, only a small time lapse accompanied, probably, by small body language signals. Coinage of the new term, self-defeat, was useful here.

Self-defeat can be highly damaging because if **A** intended bodily harm on **B** he then harms him or herself dramatically, from burning a finger to self-flagellation. A child may be angry at the cruelty of a teacher and be unable to express this and then turn the anger on self, including destructive tendencies like destroying toys.

Dealing with self-defeating, retroflecting behaviours.
F. Perls[34] said: "attention can be focused on the *self* part of reflexive verbs.[35] *Self*-reproach is turned into *object*-reproach, is turned onto the real target in the outer world." Perls, *et al.*[36] say always "reverse the direction of the retroflecting [self-defeating] act from inward to outward. The energy also reverses, is liberated from the musculature and impinges on the environment.

Appropriate questions to the person who is doing **Q** to self are –

• Who would you like to do **Q** to?
• When would you like to do the **Q**'ing?
• Where would you like to do the **Q**'ing?

Thus filling out the specific aspects concerning **Q** and **Q**'ing and changing the dissociations of retroflection to form associations.

Awareness often arises when a passive form of self-defeat is turned into an active form. The wish to self-**Q** is countered with the challenge –

• When are you going to start **Q**'ing?

Insight may be gained on seeking rationalizations for the avoided issues in self-defeating activity. The aim should be to induce awareness of the positive issues of self-realization and self-responsibility. The issues can include expression of anger, including violence (on a cushion, etc.) Awareness of self-defeating activity is promoted by acting out and this involves active extrusion. So separate the active and passive components, the top-dog and the under-dog (in Perls' terminology); the top-dog is **Q**'ing, or threatening to **Q** the under-dog.

Since self-defeating activity is internal a suitable antidote is to turn the internal direction into external and thus use the dialectical, polar opposite. Perls gives the example of turning the self-reproach of the person who feels inferior into self-support.

On the contrary, if I am about to engage with you and you sense this, and I withdraw we are both in limbo. I, certainly feel bad because I have not had the courage to even contact you. I have defeated myself. Over time I feel guilty and ashamed of myself, particularly if I meet you again. The stress on feeling is more appropriate to the situation where self-defeat spoils the interpersonal relationship.

Confluence

An obedient, unquestioning soldier conforms without question.

Introjection, projection, retroflection and deflection may be all occurring together or in close sequence for the confluent person who is gently aware that his interests are exactly the same as those of his friend. He conforms like a robot but his conforming is repellant and the recipient does not like it.

Most citizens conform to stated and unstated rules. They try to "Keep up with the Jones!" Fashions come and go but must be colluded with. Just as exacting is the situation where the person deliberately avoids anything fashionable and maintains an aloof independence. Confluence is a stuck state, is total disablement as the person becomes a slave of everyone around.

Somewhere between conformity and isolation is a state of alienation and independence where the person carefully chooses some fashionable objects and chatters with the conforming folk while assessing what is suitable for self. Rules are to some extent social graces. Good table and other social manners are pleasant for all concerned especially where hygiene is involved.

Perls' confluence is a "latching on," conforming process. The conformer likes being with whoever is near by but does not say so: "OK if I stick around a while?" He or she just hangs around, not wanting anything much, and doing everything that the other person does. The other person may feel complimented to start with because being with the conformer can feel warmly supportive, specially if from a person who counts, a "personality" in modern jargon. But after a time the receiver becomes aware of the leach effect, as energy is sucked up by the conformist.

There are healthy and unhealthy forms of this kind of activity. It is OK to conform at the moment of engagement with someone provided that one goes on to full separation in the activity stage. In self-inflicted frustration conformation indicates a failure to recognize the action necessary to continue a task which is clinging to unawareness. "The patient sees to it that nothing new will occur." Muscular paralysis prevents emergence of bodily sensations. "The aim is to get the other [person] to make all the effort." The conforming person, like the intruder, is always attempting to get another person to complete his figures.

Beneficial confluence. As the Polsters put it, the confluent, conforming person goes along with the trends of his time. They are uncritically pleased with the most recent pop music, clothing fashions and other fads among his friends are his too! He is boon to the pop industries. The confluent person is adept at avoiding risk taking and thus avoids self-nurturing. If he makes a mistake and avoids a situation requiring conformation he may feel guilty and shameful and may approach the "wronged person" with abject apologies. He may indulging in self-defeating activities expressing anger about his social situation. His stupidity may be evident to everyone but himself, and he becomes mercilessly exploited, as illustrated by Moravia in his novel about life in fascist Italy called "The Conformist."

Dealing with confluence. Beneficial experiments –

- Say hello! Say Good-bye! And shuttle between these to clarify separation.
- Promote awareness of the current style of engagement.
- Articulate what is going on.

One aspect of conformation is that the affected person expects 100% satisfaction of needs from spouse or other person. This can be explored by saying –

- I want you to be …
- I demand that you are …

- I am disappointed that you are not …
- My well being depends on you changing.
- Tell him that his shit stinks!

A conforming person marries an ideal fantasy rather than a real person.

Egotism

A confluent person who is essentially, catastrophically, dramatically self centred is known as an egotist and is totally dissociated from surrounding people.

New engagement styles

All the defensive measures of the ego against the
id are carried out silently and invisibly …
[though] we can reconstruct them in retrospect:
we can never really witness them in operation.

Anna Freud

F. and A. Freud were interested in many engagement styles whilst F. Perls was interested in the five set out above with the possible exception of egotism. Over some 30 years of practice as a gestalt therapist, mainly running groups, I came to consider that there were many more engagement styles and that each had both beneficial and deleterious aspects, as was discussed above. Some of these are discussed below and more will be elaborated elsewhere, in due course of time.

Beyond the half dozen engagement styles (defence mechanisms) set out above are the possibilities of scores more. As examples, Clark (1998, p 7) lists: "S. Freud 17, A, Freud 10, Cramer 50, Bond 24, Laughlin 22 and DSM-IV 27." It is clear that there is no universally accepted complement.

While I am interested in the universality of engagement styles I recognize that what interests me in the therapy situation are adverse events for the person concerned, events for which amelioration is sought. And there still remains interest in of beneficial associative engagement styles that the person can enjoy when the adverse one(s) are done away with.

So, in discussing an engagement style the following questions arise to enable turning repellant associations into welcome associations –

- has this style entirely beneficial consequences
- are there also possible deleterious consequences
- is this style entirely deleterious in consequences
- if deleterious what do I need to do to ameliorate the consequences.

There now follow notes on a few engagement styles set down to stimulate thoughts on the matter rather than be definitive. They derive from some 30 years as a practitioner in the gestalt style, usually in group work, and always free from restrictions that an employer might provide. In contrast consider Clark's situation (1989, p xii) as an employed student counsellor where he was aware

of denial as an important "defence mechanism" as the students lied to him about their academic attainments and in their excuses (mythology) for their predicament.

In my groups experimenters sometimes corrected a statement made on a previous time and I was not aware of "bare faced lying" so *denial* was not an important engagement style in my experience whilst one that could be designated *amending errors* was.

Before proceeding it is worth emphasizing Anna Freud's interests which were that the therapist aims to combat anxiety and foster happiness. Also Ivey, (1989, p ix) wrote

– … defences [are] a client's natural way of being, partially learned, partially innate … defence mechanisms are a logical result of developmental history of the person in environmental context.

I am concerned with set-piece events that happen in response to recurring stimuli. Examples; I am in a motor car and know there is a risk of an accident so I fasten the seat belt. If I meet a large man with a bellowing voice I feel fear and back off.

The set piece events were first described by S. Freud, reviewed and more clearly elaborated by A. Freud (1937) and considered to be defence mechanisms since it was evident that, in the pathological situation, the person was defending self against occurrence of a trauma. Such defence was sometimes of doubtful benefit as the defence mechanism adopted could entail alternative traumas.

In recent years it has become clear that these so called defence mechanisms have a much wider role than reaction to trauma. They have valuable functions in every day, healthy life. This was pointed out by Wheeler (1999) who preferred to refer to them as contact functions.

As explained on pages 4 and 52 note 42 I don't use the term contact and so adapt Wheeler's term to make engagement functions. This is still, for me, not definitive as 'function' is a mathematical term and fleeing from a tiger is not a function of a person in the way that sine can be a function of x. Action is a better term – 'engagement action' describes an active behaviour. A better term is available but needs an extended introduction. It will appear at the end of this section.

Cooperation

Venturing out in ancient times almost certainly involved association among people. It is easier to bring down a deer if three people are hunting it. And when it came to a mammoth it needed six or eight people – probably all men.

A person rarely creates a grand edifice on his own. He needs colleagues to deal with special aspects of the work, even if it means going to a shop to buy stuff. A gardener needs seeds, a wall painter needs paint, a clinical doctor needs administrators to manage the cash.

Within a family giving and receiving are natural styles of interaction and engagement and the matter only becomes remarked when someone gets in a state and refuses to cooperate. In other communities persistent denial of receiving may not be remarked as people tend not to bother with uncooperative persons.

Well planned cooperation benefits all concerned and is typical association behaviour. A rat in the system can destroy it as did an administrator who managed the accounts in such a way that he gained ownership of a doctor's clinic.

There is a political aspect of the cooperative system in which all participants contract to cooperate with one another in distributive or producers coops.

The opposite of cooperation is competition during which the master – servant relationship prevails. Cooperation and competition can function together in a community if the prevailing ethos is that of liberal capitalism.

Competition

In the capitalist system entrepreneurs compete for trade and it is "devil take the hindmost". In "good times" everyone is happy but the system is flawed and collapses periodically as is happening as I write. Then people lose their jobs, houses, spouses, are anxious, very unhappy and long for the good times to return. The bankers meanwhile make millions in cash. Competition is good for the few and can be terror for the many, especially if wars are organized to stimulate the cash flow and/or gain mineral wealth such as oil.

Competition may appear to be a cooperative activity with the workers cooperating with the boss but is soon seen to be dissociative when the boss lays off the workers.

An arrogant entrepreneur dissociates from employees and hates cooperators, considering that their caring and sharing constitutes unfair competition.

Venturing

For me to sit alone inside my skin saying "I need something-or-other" is self-defeating and does not feed babies.

I have to go out in the environment or field to find what I need. This is venturing.

In this preoccupation I am happily expecting success while tremulous about the possibility of not finding game. I may be alone or associating in a line of hunters with similar expectations.

In a non-healthy situation the person wanders out, unheeding of needs back home, captivated by his circumstances, ungrounded.

Out in the environment the person had declared a willingness to donate (give) to me what I want in exchange for something I had already or a sum of money. From their point of view, the activity was donation conditional on a monetary exchange.

Isolation

The isolated person is really stuck as he or she dissociates and avoids encounters and engagements. He or she may be brought to see a therapist who uses his skill to determine what is going on behind the isolation and facilitates the process of engagement between the two of them. Such isolation is a form of anti-confluence and dissociation.

The person may isolate him or herself from particular people and not all people. Remember Oedipus isolated from his father and bedding his mother.

Clark, (1989, p 85) writes: "Isolation as a defence mechanism enables individuals to exclude intolerable conflicted feelings …" The isolated person talks with a flat unemotional voice and tends to use rational descriptive language.

Isolated people very rarely meet facilitators.

Denial

As detailed above, Clark, (1989, p 23), had to contend with students who were referred to him for academic inadequacy but said they were satisfactory. He was primarily concerned with *denial* of the truth which occurred as the student attempted to avoid expression of "intolerable affect and conflict [inducing] rigidity and stereotypical behaviour". Clark described varieties of denial:
- denial of the seriousness of a situation,
- denial of personal relevance and responsibility,
- denial of urgency, denial of relevance of affect,
- denial of relevance of threatening information.

He also described ameliorative procedures, changing dissociation to form association, for the therapist to use; reflect back Rogerian style, explore interpretations, reformulate (reframe), using 'stop' tactics (page 167), and act into the contrary possibilities.

One can expect *denial* to occur in other communities where experimenters do not express free will when meeting the facilitator.

To emphasize; *denial* is closely related to falsification, to lying.

Anticipation and Forestalling

The experimenter who finds him/herself in a situation of intense dissociation and criticism may choose to follow a course of anticipating criticism and behaving in such a

Driving fears

X came to facilitation group complaining of "morbid fears" while she was learning to drive a car. The instructor encouraged her to use her speed appropriately but in the test the examiner failed her and complained that she proceeded too slowly when the conditions were right for going faster.

X He said by being too slow I made the drivers behind me go slow and that would frustrate them and make them angry.

F So you were not concerned about the feelings of other drivers on the road.

X No. Damn them. I look after myself.

F You look after yourself by disregarding the effect you have on your colleagues on the road. Don't you expect that the antagonism you engender might be much to your disadvantage?

X How do you mean?

F There could be a young man in a Merc' behind you fuming because he wants to get past you and get to a business appointment on time. Then, when he overtakes you, he viciously cuts in on you, causing you to brake and you get angry with him. Is that a way to look after yourself and a safe way to drive?

X [Quietly] Something like that happened.

F What will you do in the future?

X [Vigorously] You don't understand! I get scared if I exceed 30. I tremble and get worried about my steering and get muddled about braking and accelerating.

F I see that you are trembling now. You can slow down and breath more deeply.
[pause] Was there a time, perhaps long ago, when going fast was really bad for you?

X [pause] There was a time. I was pushing my dolly in her pram but the pram got away and crashed into a car.

F [pause] Dolly in the pram hit a car. What happened next?

X Dolly was terribly upset. I had to hug her to stop her crying. Mother hugged me and said that nothing was damaged because the car wasn't moving. But Dolly was upset even if she was not hurt.

F Can you put Dolly on the cushion there, and talk to her.

X [Put her handkerchief on the cushion] You are not hurt. There is no need to cry. [X grabbed the handkerchief and hugged it to her breast. She trembled and wept while muttering] Don't be scared. Not hurt, not hurt."

F [Sat silently until **X** had calmed down]. Instead of being scared, what could Dolly be?

X Dolly likes to have gumption. She gets on with things without fuss and bother.

F What is gumption? I've not met that word before.

X It's what my Mum said and she meant have courage, be bold.

F Thank you for the new word. What is happening for you now?

X I feel different.

F What is different for you?

X Well, like, I came in worried about driving and now I'm interested in lunch.

F I have a suggestion for you.

X What is it?

F When you go driving be sure to hold a handkerchief and tuck it in your pocket as you settle in your seat. The handkerchief means courage for you so that you can go as fast as necessary and be happy with the other drivers on the road.

F and X met on following weeks and X reported that her instructor was happy with her progress.

F was tempted to open the session by challenging X concerning her fears but chose instead to establish Rogerian rapport with the belief that X would present her affect on an appropriate moment and that was how things turned out.

Reformulating "scared" to "courageous" helped Dolly to see her world in a different way. Dolly couldn't go out in the car but was evidently happy that handkerchief could represent her interests. And, as usual, X emulated Dolly.

In terms of engagement styles, X was relating strongly to her Dolly and only secondarily to the facilitator. This was not important because X projected first anxiety and later courage on Dolly which she then introjected. The latter was assimilated helped by expression of affect. Later she acknowledged that her engagement style had changed from fear of speed to hunger for food – without using gestalt speak.

way as to forestall the possible consequences of default. "I hate criticism so I will behave in such a way as to avoid criticism."

This is a **very** common maneuver: "I hate cheese but I eat it so that she won't go on at me about it." "I don't want to go to hell so I'll be a good church goer."

Catastrophic expectations

This person has a fear of getting things wrong and usually, when endeavouring to do something, it does go wrong. The central impression that something terrible will happen prevents him/her from thinking about anything else.

The person tied to this engagement style has a generally negative view of everything and needs support to discover the value of positivity. 'What is the worse thing that could happen for you?'

Aggression, assertion and passivity

Civilized people can cooperate with a very mildly dissociating, aggressive person but shy off engagement, if they can, with red-faced aggression. So a person who offers aggression as an engagement style will be spoiling his relationships with other people and needs facilitation to restore equanimity. However, aggression has many little brothers, techyness, annoyance, and a big brother in violence.

A civilized person may decide to not bother with a passive person who always relies on other people who tend to react away to avoid being exploited.

The aggressive person tends to also be an authoritarian provocateur and tries to get other people involved in being a nuisance. The authoritarian employer has a hold o ver people on his payroll. Such one-upmanship, a game of ascendancy played by managerial staff over juniors, is more civilized than fisticuffs. So are Berne's games that people play, a further development of the Potter[38] attitude

People tend to react to an aggressive person by becoming passive while a passive person wonders why the people round him or her become aggressive. Aggression and passivity, mildly exerted, may get a person what he or she wants without generating an extreme reaction from other people but the safest way of behaving is being assertive.

The polarity here is expressible verbally, as in the heading of this section with assertion as a generally acceptable, associative median state. Both aggressive and passive people are helped by meditating on the theme of assertion and conducting experiments in group so that they assess their effectiveness for themselves.

Reality checking

When events are happening at great speed it is often a good idea to

STOP

breath easily, and ask self and/or others:

"What's going on?" and so seek association.

This tactic is particularly valuable when a degree of panic sets in or when a lot of people are all talking at once, as in a "brain storming session."

Displacement

Clark[37] (p 63) describes displacement as "… redirecting an emotional response to a vulnerable substitute." Wives often suffer after hubby has had a bad day. Children come home from school and take their frustration out on little brother or the cat.

Identification

Clark[37] (p 65) describes how a vulnerable person tends to solve problems by copying the behaviour of people who have already solved the problem. This is part of the education system, of course, but a member of staff promoted to manager eases his way by remembering the ways of his predecessor.

A boy who is beaten by his father tends to beat his sons when he is older.

Home

"Home is where the heart is", so my mother said. Home is my special cave where the fire is always warm, food is always available, where, circumstances permitting, I can copulate and where I can relax and sleep in peace.

Home is where my ideas live, where my fundamental beliefs are not challenged and I find what I like most.

Home is where I protect my territory (NIMBYism), expel non-cooperating intruders, and am happy and without anxiety. Home is where I associate with and welcome my friends.

Love

The 'being in love' syndrome has me in two conflicting states. I am either with my beloved and calmly content or without my beloved assailed by aches and pains. Thinking about the beloved tends to displace all other thoughts. Teenagers are particularly prone to such love-lorn, confused living styles. Love is perfect association.

Love, requited or rejected, is the subject of great art, novels, operas, paintings, sculpture.

Hate is the most dissociating, destructive emotion.

Potentiality

A healthy person has all engagement styles available for action. The unhealthy person is thinking about something else – a failed projection or two – and does not remember engagement styles at appropriate times.

Suffering

There must be an engagement style term for the long suffering person. A main characteristic is self denial which shows in a hang-dog face and drooped shoulders. This is often a woman who, though middle aged, is a slavey for her mother.

Polarities and experiments

We are mostly concerned here with events that occur and constitute problems, from embarrasment to anger. If the person can control events there is no problem. Otherwise the person is at least confused and not knowing clearly what is going on. It is up to the therapist to study the situation, see what is going on and suggest experiments for the affected person so that he or she discovers what is happening and finds a solution to the problem for him or her self.

The basis of the experiment can be exposure of the person to the polar opposite situation to that of the signs and symptoms of the prevailing engagement style. Thus if the person is afraid to venture out of his house the appropriate experiment would be to fantasize about the consequences of very low key venturing out – as is the usual "treatment."

The principle behind polarity work is that all polarities have an equilibrium, median state. For aggression and passivity, as noted above, resolution produces the assertion state from which one can see both aggression and passivity and choose to use them if needed. The experiment for the person would be based on exploration of the opposite pole to that usually encountered.

However, many polarities do not have a named equilibrium, mean, median state. It is necessary then to pretend that there is one and the experimenter in talking about it may speak voluminously. An example here would be the outcome of consideration of *venturing out* and *staying put*. The experimenter may talk about feeling poised and ready to either venture out or remain in; I know no one word that clearly expresses this but that is no excuse for avoiding the issue. A rather silly sounding word is *teetering*; teetering on a balance point or boundary ("sitting on the fence" unable to decide what to do; procrastination – an engagement style in its own right).

Odds and ends

The expression of slogans can indicate that a defence mechanism is active, for example saying: "He is knee high to a grasshopper" instead of saying: "He is small." The expression here is to do with grandiloquence and poesy.

There was time when some folk had a stock of latin tags that they produced when appropriate and to make an impression. I am inclined to produce bits of Hindi – evidently from a very impressionable time in my life, late teens.

Do jokes indicate the operation of a defence mechanism?

Personality traits

Academic evolutionary psychologists have recently given attention to categories of the character of healthy people. LaFrenière[39] writes –

> Personality systems may be thought of as evolved systems designed to deal with life's dangers and opportunities. Research in personality has revealed five personality trait dimensions – the five factor model of personality (FFM). The "big five" personality dimensions are usually named for one end of the continuum they represent [… and these are]: extroverted… , agreeable… , conscientious… , stable and open… .

Personality traits are partly inherited and partly generated in contact with the environment in which the child grows. LaFrenière points out that the personality characteristics of babies and infants, usualy referred to as temperament, correlate with the personality traits that are recognized later in life. So temperament and its failures are also relevant when considering interruption processes.

Temperaments and personality are not absolute characteristics but are subject to spread on the continuum of the familiar bell-curve with most of the population in the middle range and a few at the extremes. Each characteristic is inherited with the observed manifestations changing with time, as the child and person grows.

On reversing these phenotypic characteristics we have a series of interruption processes; introverted, disagreable, dilitante, unstable and closed; none of which are congruent with the interruption processes that F. Perls, A. Freud and G. Wheeler favoured or the others I have added. Nevertheless, we will need to consider and integrate these personality traits in due course of time.

The effect of occurrence of multiple engagement styles

I'm inclined to be interested in happiness. "Is so-and-so really happy!" I have to admit, though, that a very common human state is of a sort of neutrality where there is no sign of affect but which seems to be more akin to happiness than anything else.

People indulging in beneficial introjection and projection can be very happy indeed and it need hardly be stressed that happiness is very common among civilized, intellectual people!

A competitive person readily becomes annoyed if anyone gets in his or her way.

A deflector may show his guilt by reversion to passivity.

A conformer may become very angry if challenged about his lack of self-confidence and self-assertion.

A cooperator may become annoyed if a supposedly cooperating colleague does not pull his or her weight.

Somewhere between being intrusive and extrusive is the blemish free state of venturing. As was made clear above, extrusion and intrusion may be related if the material concerned was common to both processes.

Notes and References

1   Anna Freud and F Perls and colleagues referred to *contact* and *contact functions*. As explained on pages 4 and 52 note 42 the word *engagement* is a more precise description of what happens between people than is *contact. When people associate.*

2   Anna Freud, 1937

3   Perls, Hefferline and Goodman, 1951.

4   Polster and Polster, 1973.

5   Wheeler, G. 1991. Wheeler refers to engagement functions and I prefer to refer to styles of engagement or engagement styles. I am suspicious of the term function as, in mathematics, it refers to a derivative of the subject matter. Thus, if $y = \sin x$, the sine transformation is the function of x that determines the value of y. In examining engagement styles there is no sense that anything is derived from something. The style is what happens, how the module of nerve tissue is manifest in action. There is no sense of a transforming function between the module and the event.

6   People tend to be more excited and less inhibited with their animals – cats, dogs, horses – than with other people, even supposedly close ones, except perhaps neonates.

7   Swanson, 1980, p. 76.

8   Perls, F. 1947, p. 45.

9   Perls, Hefferline and Goodman, 1951, p. 222.

10   Perls, F. 1969 & 1974.

11   Perls, F.1969, p. 27.

12   Perls, F. 1951, p. 218.

13   Polster and Polster, 1973, p. 89.

14   ibid, p. 91.

15   Perls, Hefferline and Goodman, 1951, p. 248.

16   McCabe, p. 48.

17   Swanson, 1980, p. 73.

18   Perls, F. 1947, p. 237.

19   Perls, F. 1975, p. 115.

20   Perls, F. 1947 p. 244.

21   ibid, p. 255.

22   ibid, p. 256.

23   ibid, p. 257.

24   Perls, F. 1973, p. 81.

25   ibid, p. 82.

26   Perls, F. 1947, p. 245.

27   Perls, F. 1966b.

28   ibid, 1951, p 528.

29   Perls, F. 1975, p. 187.

30   Perls, Hefferline and Goodman, 1951, p. 171.

31   Marcus, 1979, p. 42.

32   Polster and Polster, 1973, p. 84.

33   Perls, Hefferline and Goodman, 1951, pp. 171 & 176.

34   Perls, F. 1948, p. 67.

35   Some gestalt commentators have assumed that all reflexive words, starting with self- in the English language, indicate deleterious retroflection. This is erroneous as, for example, self-nurturing is essential for life.

36   Perls, Hefferline and Goodman, 1951, p. 173.

37   Clark, 1998.

38   Potter, 1950.

39   LaFrenière, P. 2010, page 65.

—=(☆)=—

Chapter 5

Tertiary and other Failures of Interpersonal Engagement

It's a great big shame
And if she belonged to me
I'd let her know who's who.
Isn't it a pity that the likes of 'er
Should put upon the like of 'im!

Old music hall song

Tertiary interruptions of gestalt development of the style of blame and resentments were described in the Introduction as social phenomena: for example "You should be ashamed of yourself" said aggressively. They are very much manifestations of an alienated state of mind, are dissociated states and are accompanied by a selection from a list of deleterious affects, anger, sadness, resentments, low or high energy, etc.

The experimenter is overwhelmed by interest in the game that he or she is playing and has no consistent and persistent view of any centred self activity, although these may flash into mind [consciousness, awareness, attention] from time to time. One of the tasks of the facilitator is to notice and promote awareness of these flashes of centred self activities so that the experimenter can know them.

Anna Freud (box on page 71) pointed out that attempts at self therapy often lead to a worsened situation. The person who is fed up with his or her shyness discovers, with the aid of friends, that cannabis smoking solves the problem. Then the weak arm of the law takes the weed away and leaves the victim lonely and bereft.

In the following discussion of the alien games the pattern of approach will be –

1 description of the game
2 components of the game
3 ameliorative steps that may be taken when dealing with the game.

However, the games are not autonomous. Each has overlap events with others and it could be possible to present the catalogue in many different sequences. I choose to start with blame because, in human behaviour, it can lead to the others as the blamer tried to make other people take responsibility for what is going on.

Alien "games"

Blame
Shame
Guilt
Victim
Persecutor
Rescuer
Resentment
Vengefulness
Self-pity
Rebelliousness
Depression
Stress
Hypochondria
Phobias

Blame

I reckon a chap who can smile
when things go wrong
has thought of somebody to blame it on.
Old Yorkshire saying.

Yes! When something goes wrong dissociated self seeks someone to blame. Centred self may analyse what is wrong and puts it right, seeking help if necessary. If centred self has caused the error he or she admits the fact, makes restitution and without emotional contamination.

Blaming others is a popular children's game. "I didn't break it! She did!" Adults become more sophisticated – the leader of the opposition blames the Prime Minister for the global credit crash.

Blaming self morphs into shame and guilt. Jump down four paragraphs to find guilt and F. Perls' amelioration procedure.

Shame

Persistent blaming of self sets up shame, a complex of ruminations over whatever went wrong and negative feelings, anger at self "for being so stupid", sadness at "having done it again", low energy, dissociation from self-interest, little interest in anything else going on and a tendency to pass the blame to other people, "shame on you!" no matter how apparently innocent.

I feel ashamed when I fail in living up to other people's expectations. Some kind of contract was involved – I agreed to do something and failed. This happens in the best regulated families but instead of courageously saying "I did the best I could under the circumstances" I skulk off, hide and assault myself with reproaches in my implosion, at my special impasse. I stay in healthy figure if I confess and this may be an altered figure but at least it is not a meander off to the confusions of the impasse. Sulking and ashamed, quiet, hiding, is teetering on the edge of the impasse. Self-responsibility leads to straight talking and OKness or retreat leads to self-ostracism.

The word "shame" is particularly associated with blame and the word "should.": "You should be ashamed of yourself" thus revealing the aggressive, projective character of the situation for the sufferer. Though shame as an aspect of character is very much a passive state: "I should be ashamed of myself and I will hide away from people." The "should" aspect reveals the introjective origin of shame and projective activity "You should be ashamed," can be compounded with retroflection: "I can't tell him he should be ashamed so I should be ashamed." Such rubbish is often accompanied by "Sorry!" "please excuse me" or "I'm afraid that …" and other weak attempts at connection with the other person.

A helpful suggestion for people who wallow in shame, self-contempt, self-disgrace, self-defiling, self-embarrassment, self-humbling, self-ignominy, self-mortification and/or self-pity is to invite the experiment and break out of the repression by beginning phrases with either "the opposite of shame for me is …" or "I think I should have done … at such and such a time." The invitation then is to abandon passivity probably leading to expression of aggressive intent, leading to resolution of the polarity. The client comes to see the value of shame as self-protection and for dominating other people and had the opportunity in the future to choose among the available behavioural options. The very old Cockney song, quoted at the head of this chapter, expresses the ideas if this section very well.

The person blocked off with shame benefits from two chair analysis of what is going on. The facilitator helps the experimenter to choose contrasting aspects, perhaps the benefits of passing on the shame versus the disadvantages.

Mollon[2] stresses the importance of the role of shame in perpetuation of the dissociation in multiple personality disorder. Shame is initially projected from abuser to victim who introjects it and then repeats it. Shame is a self-sustaining, circuitous phenomenon; the shamed person becomes ashamed about his sense of shame. "The shame of it!" See the Ivanov Box.

Ivanov is speaking: It's an extraordinary irony. I used to think and work all the time and never felt tired. Now I do nothing and I'm completely exhausted. And all because of my conscience. Hour after hour, eating away at me. All the time I feel guilty. But of what? … I look round, I have no money, my wife is ill, my day goes by in constant meaningless gossiping and squabbling.

How spent I am, how I despise myself. It's only a drunk … who can still respect me. I hear my own voice and I hate it … I search for faith, I spend days and nights in idleness, in doing-nothingness, my mind, my body in permanent revolt. I look out of the window: my estate is in ruin and my forests are under the axe. … I have no hope, no expectation. My sense of tomorrow is gone … the shame of it … What has happened to my strength?

Guilt

A person blaming self for wrong doing, feeling ashamed knows a vindictive alienated state and says something like: "You are guilty and should be punished." It makes little difference whether the experimenter is really guilty or not.

This is an opportunity for theatricals. The role play can be poor little victim versus a bewigged and berobed judge. If the therapy room has some props, all the better.

Guilt and shame indicate the occurrence of pungent problems and exemplify the occurrence of bad conscience, multiple avoidance phenomena and loss of self-nurturing. They have already turned up in several of the categories set out above. **Jock** said: They think I should be different so I feel guilty about not obeying them and I am ashamed of my indolence.

Since expressions of anger are loaded with aggressive exaggeration and verbal generalizations, nice people tend to take avoiding action. This results in residual, unexpressed weak resentments and feeling the negative emotions guilt and shame. Guilt is associated with "It's my fault" and "I should [not] have done it" and shame with "I ought not to have done it!" The "cure" for psycho-asphyxiation by either or both is to find and express the resentments and eventually the ill-expressed anger – it often comes up as "I feel rather upset (tetchy) about . . ." – and other weak expressions.

Forms of self-reproach, possibly retroflective, lead to feelings of mild guilt, a kind of pseudo-guilt since the major affective reaction: "How disastrous!" is not usually extant. Some pseudo-guilt often occurs as a result of internal chatter: "Have I made a mistake?" "A confluence of introjects produces guilt."

F. Perls recommended discovery and expression of the resentments that he considered always lay beneath guilt and on those premises the same is true of shame. The trouble here is that the client may resent the suggestion of the facilitator thus diverting attention from the original presentation and compounding the problem. An exploration of polarities successfully leads the client to find the resentments for him or herself and can choose to deal with them or not.

Mistakes, blame and shame. It is idealistic to expect all gestalt figures to proceed perfectly to maturity. Mistakes occur from time to time and the important aspect is how the mistaker reacts to the event. A totally healthy person says: "OK, so I've made a booby – what do I need to do to get things right again." And it may be cooperative with his colleagues if he says "I'm sorry but I was doing my best."

Other people get angry with themselves. "What a shame to waste this much time and effort!" They may turn to others: "It's your fault. You should have been ready." In either case the familiar figural residue and bad feelings eventually go blank, impasse and implode, settling down as alienated behaviour.

The victim

Blame, shame and guilt load a chap down. He or she soon feels that all the world is against him or her and whinges on about it for everyone to hear. One of his triumphs is to obtain attention from other people and these come in two sizes, a rescuer may want to solve all the problems "There now. What can I do for you, poor thing." The larger size is a persecutor who will load blame on the victim. They form an alliance, batterer and battered, and woe betide the rescuer because both will turn on him for interfering when it is not his business.

A **victim stance** requires relationship with a **persecutor** and hope for a **rescuer** as Berne[3] neatly demonstrated. Switching among these roles sometimes appears to be humorous to an uninvolved observer.

The persecutor

This aggressive person spends his or her life looking for victims to lord it over. He or she may have assigned roles, from lord of the manor in medieval times to blue jacketed police thug. And he or she already occupies a prominent position in the above account, in alliance with the victim.

The Rescuer

To develop the notes set above, Berne sketched all this out neatly is his drama triangle where the victim, persecutor and rescuer dance inanely in a circle. I remember two colleagues, each with a strong local dialect, one Mancunian, the other Australian, arguing fiercely. The matter was settled by an observer, me, who, after enjoying the fun for a few minutes, pointed out that the key word in the argument was "rations" for one and "Russians" for the other. Otherwise, chums, attempts at rescue, to form associations with others, readily lead to dissociation. The solution is provided by the observer aspect of self who stands back and tells the others what is going on because they usually have no idea of the ridiculous state that they are in. The observer, is of course an aspect of core self who appears later in this chapter.

The resentments and grudges: suppuration and poison

A citizen who perpetually finds life populated by people who appear to do him or her down becomes preoccupied with resentments. He or she may talk about feeling bitter, wanting to blame anybody and everybody and self.

Something has gone wrong and the experimenter has reacted in a self-protective, self-defeating way. The experimenter needs to express and share the bitterness, anger and sadness felt rather than keep it in.

Unexpressed resentments are indeed toxic – the need to express whatever is begrudged, repeatedly surfaces and will continue to do so until expressed, either for real or in fantasy therapy. Behind the repression lies fear: "What will happen if I am so bold?" The opposite of "I resent" is "I appreciate." Awareness and expression of these poles helps restitution and the resentful ghost can be exorcized and laid to rest. F. Perls[4] calls the process of retaining the resentment "The hanging on bite."

Neutrality

This is a stance taken by deflectors who like to pretend that they are above, dissociated from, what is going on. A person who openly declares neutrality is not deflecting.

Vengefulness

This person takes matters actively towards taking vengeance, serrupticious or violent, on whoever he feels is to blame for something that aggravates him. The extreme examples are the vendettas that occur when a family member feels obliged to kill a member of another family, a member of which killed the vengeful persons family member. Such vendettas, shared among a group of people, can be cata-strophically, murderously destructive and go on over long historical periods. A sickening progress of foolishness.

Introspection

To be aware of potentialities
is to be fore armed

Introspection is deprecated by Perls, *et al.*[5] because they consider that part of the self is split off from the part observed but this can in fact be a valuable self-evaluation and an awareness learning process as was set out above.

Criticisms and Comments

Starting from the premise that civilized people, on a feeling level, do not hurt one another, do not hurt self and do not accept hurt to themselves from others, and that, on a reasoning level, a comment is either true or false, comments and criticisms can be connected by an algorithm such as that of Chart 5·1. As a comment is a gift to another person, criticism is a comment plus negative encumbrances which hinder acceptance of the comment part. Further analysis enables separation of the hurtful aspect of criticism from the valuable part. In terms of the figural events. Comments are attempts to nurture the other person. Criticisms are power play that put the other person down.

This is all in aid of running clean figures. The generalized self-nurturing cycle, *Frontispiece*, shows clearly how making a straight forward comment produces an outcome that is satisfactory to both people while negativity of the criticism has the potential for pushing the receiver out of his or her organismic cycle. In this depiction I envisage that **Y**, when criticising, will effectively jump over the polite stages, <3> and <4>, and blunder onto stage <5> with a heavy weight attack. **Z** is then devoting more energy and attention to his or her own anger (or annoyance, etc., for less reactive people) than to the information offered in the comment component.

Before going on I want to divert attention to the dictionary definitions of criticism. There are two basic meanings and the one that does not concern us here relates to the evaluation of books, musical performances, etc. Of the other meaning the Oxford and the Collins dictionaries say; to pronounce judgement, to censure, to find fault. Chamber's Thesaurus gives as synonyms; reprove, censure, reprimand, cavil, fault-find, quibble, blame, detract, disparage, be sarcastic. There is no way that to criticize is synonymous with to comment, as I have found that many students and group members believe. Criticism is aggressive power play and the outcome depends on how sensitive the receiver is. A dullard will just smile and accept it. An aggressive person will become angry: "Who do you think you are talking to!" An assertive person is pulled up short while thinking about how to deal with this aggressive attack: "I notice that you are criticising me aggressively and I don't like it. If you want my co-operation be more gentle with your comments and suggestions."

Returning to Chart 5·1 I can say that I have set it out to represent what I consider a centred, assertive person with good self-esteem would do in dealing with comments, some of which may turn out to be criticisms. On receiving a comment there are two basic questions; "Is this true for me and if not I reject it" and "Is the comment presented in a kind way or with malice." In the latter case I express my feelings while rejecting the insinuation.

In general I recognize that anyone can make mistakes in relationship to what he or the other person considers to be the truth and can reconsider the event on being challenged. This is effective if done gently and with good humour. My intention is to avoid malevolent replying in kind – it is standard practice in some communities to trade insults, tit for tat, with no benefit to anyone unless it is clear that good humoured, jokey jousting is intended. For me the most important action is to take action immediately to assert my case, true or false, kind or unkind. While doing so I am myself careful to avoid projecting bad feelings onto the other person; no anger, jealousy, competitiveness, guilt or shame.

In summary, if this paradigm of approach to comment and criticism is valid, then the true and kindly comment can be taken as information, pleasantly presented and agreeably accepted. It is best prefixed with something like: "I have an idea about what is going on now – would you like to hear it?" The recipient can then acknowledge the gift – "What you

say is true, I will consider doing something completely different" – and then give thanks to the commentator for bothering about it at all. If the comment is true and presented with spite the recipient can acknowledge his or her rising bad feelings – "I agree with what you have said but I don't like the way you said it. What do you gain by being aggressive with me?" If the information given was pleasantly presented and not true both partners can feel satisfied and the recipient can thank the presenter while gently asserting the falseness of the proposition. If the information was presented maliciously and is wrong the receiver tends to feel bad – "You are wrong *and* I don't like the way you are talking to me and if you are interested I will tell you what I think is right. Perhaps you could reconsider what you are on about and tell me in a more respectful manner." There are several ways of reaction by the receiver that are not helpful, using false politeness, saying "Sorry …," for example, "Excuse me …." or "I'm afraid that …."

I started with the premise that we were considering the interaction of two people. This can be extended to self-criticism and self-commenting. There is no advantage in lying to oneself. There is no benefit in hurting oneself. So I treat myself in a kindly, gentle, detached, non-hurtful and balanced way and watch my feelings, a sure signal of right and wrong in this situation. I am tolerant of my mistakes – I am OK with making a mistake once, to do so twice is a bit foolish and there is something really wrong if I make the same fault three times. If I find myself criticising myself I sense that an inner conflict is emerging, perhaps my alien-self is having a go at my core-self or perhaps some other polarity is exerting itself. A two-chair self-dialogue is a very creative next move.

Sarcasm is a particularly caustic form of criticism: **X** is looking at his own writing and having difficulty reading it. **Y** says: "What wonderful copper plate you produce!"

Feed back. In the group therapy situation members give one another comments rather than criticism. This feed back forms a model for a basic approach to other people applicable in any social situation. Beneficial feed back is a gift to the listener. Positive feed back is doubly helpful, to the giver who says it and to the recipient who hears it. If there are several things to say, start with improvements you want to suggest and end with praise: "I had to strain my ears to hear what you said and I want you to speak louder. I loved the sparkle in your eyes!" In general, be specific rather than generalize, be descriptive rather than evaluative, be encouraging rather than judgmental, be positive rather than negative, refer to what you were aware of, refer to what you saw, heard, felt in terms of affect, and to sensations in your body. Also look at the person who is to receive feed back, if standing keep feet firmly on the ground, if sitting be relaxed but not laid back. Breath steadily, modulate voice tone in harmony with the words being used, clearly differentiate-off what is imagined and fantasized and most important of all own what is said with "I."

Lying. Giving false explanations or suggestions is manipulative and dissociative. A lie may appear to be self-protective when presented but soon suppurates into bad feelings.

Chart 5·1. The connection between comment and criticism

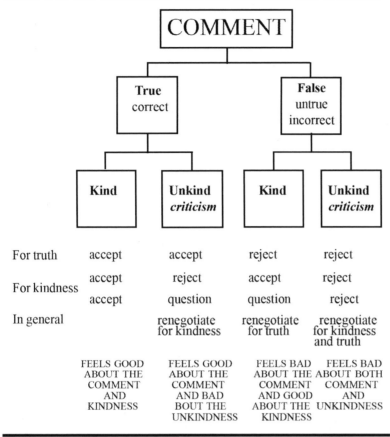

For truth	accept	accept	reject	reject
For kindness	accept	reject	accept	reject
In general	accept	question	question	reject
		renegotiate for kindness	renegotiate for truth	renegotiate for kindness and truth
	FEELS GOOD ABOUT THE COMMENT AND KINDNESS	FEELS GOOD ABOUT THE COMMENT AND BAD BOUT THE UNKINDNESS	FEELS BAD ABOUT THE COMMENT AND GOOD ABOUT THE KINDNESS	FEELS BAD ABOUT BOTH COMMENT AND UNKINDNESS

Chronic Inactivity

Adler[6] said: "in cases of neurosis we are dealing with comparatively less active individuals who naturally ... by the lack of activity ... [were unable to obtain] ... correct solutions of their problems." Sebastian Flyte as depicted by Waugh in *Brideshead Revisited* is typical of Adler's description – having been pampered when a young lordling he is incapable of helping himself. His chum, however, Charles Rider, is by contrast depicted as entirely self-sufficient and successful. (I wonder if Waugh read Adler).

Submissivity

Lack of self-esteem, like confluence, is so ingrained, that the person protects self by continually, openly submitting to the will of others. Dickens depicted Uriah Heep as just such a character

Contrast Adler's: "the egocentric striving for personal superiority" with Satir's flow – the flow towards figure completion. This is exaggeration, in a move towards grandiosity. In gestalt terms flow has turned into forced labour.

Self Pity

When things continually go wrong a person may immobilize him or herself by wallowing in self pity. "Why does it always happen to me?" Well, it doesn't and a two-chair role play of the polarity reassures the experimenter that, first a few things go right and later that more things go right than the experimenter realized.

Self-pity and self-punishment arise as a result of making excuses for oneself after making mistakes. Self is divided against itself; an active retroflector is impinging on a passive, object part of self.

There is in all this, as Adler said, use of superiority moves to hide the innate feeling of inferiority. This is the person's internal logic that is driving, projecting rubbish on to other people and retroflecting, accusing self of interfering, so reinforcing the lack of self-esteem, and generating guilt and shame feelings.

Rebelliousness

This is a "bloody minded" person who lashes out at everything he or she dislikes. Rebellion contrasts with revolution. The latter is overloaded with principles and reasons whilst the former is all self-centred emotion.

Exaggeration

Pretending that something is larger than it really is can sometimes be expressed openly and beneficially. **Jonah**'s mother frequently told him as a boy that he would be Prime Minister, or at least of Cabinet rank. **Jonah** attended public school, university (Oxbridge) and a Guards regiment before cracking up under the load. On the other hand a nice child separates himself from parental viruses. **Kirk** was intended for the Church. He took to chemistry and physics, which his parents did not understand, and so escaped from the theology that was thrust at him at home and at school.

Pseudo-engagement

When a person says –

> I think that ...
> I feel that ...
> I mean that ...

without real engagement, the person is talking to self, is talking about thinking rather then actually thinking, is talking about feeling without actually feeling and knowing his or her affect, is talking about meaning, or intention, without actually meaning or intending anything.

The person is alienated and dissociated from colleagues and is acting into a kind of pseudo-engagement using imitation ego-language (F. Perls[7]) and this is an avoidance mechanism, a form of deflection from real engagement. It is not confluence since it is an internally focused event and the boundary is intact and rigid. It is not introjection because nothing has come from outside, it is not projection since nothing really goes out. It is not retroflection because anything aimed at the outside world was only aimed and not released and thus cannot be brought back again.

Pseudo-engagement is a kind of dummy projection in which something unreal was aimed at the world and retained at the last minute. In this sense the "thought about" is recognized as unreal and is prevented. If the unreal projection is considered to have occurred then it is re-introjected, a pseudo-retroflection. It is pseudo-confluence in the same sense that the attention of the receiving person has been held and not honoured by giving something, a trance induction statement.

Thus: "I feel that I am quite warm in my relationship with you" sucks the receiving person in to making the warm approach; a real contact statement is; "I like you." Instead of saying: "I mean to say . . ." "I say what I mean."[8]

Pseudo-engagement can be regarded as an expression of an alienated self, a way of looking at avoidance of the risk of real engagement, on a social level and parallels avoidance levels as known on the psychological level, confluence, etc.

Trauma and memory

Each man kills the things he loves
Oscar Wilde

Pavlov demonstrated that occurrence of a thunderstorm wiped out the conditioned reflexes he established in dogs. This happens to people too. **Emily** has a statement to make as treasurer at a public meeting on a Friday. Her dog had died the previous Wednesday. On being called to speak she had forgotten her obligation and what she had agreed to say. Fortunately neighbours prompted her and she said all that was necessary, though in rather a muddle. People recovering from trauma may benefit if reminded of things forgotten.

Depression

I'm not dramatizing;
everybody else is depressed.
Viva in Warhol's film *Chelsea Hotel.*

This task of making this mini-review is much easier since the publication of the excellent meta-analysis by Antonuccio, DeNelsky and Danton[9] who showed that cognitive behaviour therapy in treating depression was more effective than use of anti-depressant medicines. In addition the relapse rate was far greater for drug treatment. The authors suggest that: "Psychotherapy … should be the first line treatment for depression even when severe." I believe that gestalt therapy is at least as effective as cognitive behaviour therapy and

await a thorough going evaluation by someone else to demonstrate this as fact. I also want to acknowledge the strong influence of Gut[10] on my personal understanding of depression.[11, 12]

I like to rest parts of my case on great literary figures. In this context the quotation from young Anton Chekov's description of his character Ivanov as set out in the box on page 84. Ivanov, a doctor character, is weighed down by his thoughts, is unable to pull himself together, has long lost the taste for life and is unable understand the meaning of life. The play is devoted to Ivanov's attempts to understand himself and find a reason for living. (Hare[13]). The play was not popular in Chekov's day and surely a modern audience has no doubt but that Ivanov was depressed.

Productive and non-productive depression

Self-importance is the motivating force
for every attack of melancholy.
Castaneda[14]

People described to me the insidious onset of depression over weeks or months. Each described the mood as something like "sinking" or "blackening." Everyone has changes of mood but none so drastic as those who sink into this life-effacing lethargy and lack of involvement. I have found it important to share this information with people who describe themselves as depressed because they usually think that other people are always on an even keel. This is partly because they remember their pre-depressed life state as being generally stable. Likewise it helps patients described as manic – depressive if they understand that we all have our ups and downs and that they have a greater swing than most of us.

The basis for understanding depression[10] as a state of existence for a person and as a word is acknowledgement that depression is a condition of diminished vitality, of moderated emotional responses and restricted cognition ($<7 \pm 2$)[15]. Episodes of depression belong in a spectrum where mood is moderately energized affect in contrast to an emotional situation where outright anger, etc., are expressed. As Gut[10] pointed out anger is joined by grief and sadness, guilt and shame, disgust and hate, and, her particular contribution, by depression. These all contrast to joy and tenderness and other positive states. We thus have a group of withdrawal and avoidance syndromes contrasted with dissociative ones.[16] The polarity here is anti- and pro-attachment.

Depression then, in Gut's account, is avoidance of involvement, usually in something that has gone wrong, that is not successful. This reaction to circumstances can be beneficial. Withdrawal to non-involvement is withdrawal to the fertile void, the impasse, and in healthy people is rest, repose and relaxation followed by renewed vigour and activity. This is the state described by Gut as the "basic depressed response" and the "productive depressed reaction" or "productive depression" for short. Gut says that productive depression leads to "useful learning or maturation ... some behaviour has been reorganized, some plan revised, so that ... we function more effectively."

This contrasts with Gut's "unproductive depressed reaction" or "unproductive depression" which I like to think of as non-productive depression. Here withdrawal continues preoccupying attention, a barren, infertile void and impasse from which there [seems to be] no escape.

Reaction to circumstances is thus not always productive of excitement. As Gut[10] says depression is: "a feeling of low spirit, with slowing down and withdrawal of interest from some or many of our usual pursuits and commitments ... [T]his may be accompanied by disturbances of physiological function such as sleep, appetite and sexual desire."

In the terminology I used in Chapter 3, centred self is O.K. and can thrive in temporary withdrawal and emerge reactivated into vigorous self-nurture, whereas dissociated self inhabits the barren ground with emergence only with the effect of confirming lack of self-benefit.

One of the reasons for the ambiguity in the process of the diagnosis of depression is that the manifestations of depression are side effects of many clinical diseases including endocrine disorders (Cushings syndrome, complications of pregnancy), viral infections (glandular fever), substance abuse problems (drugs, alcohol and eating of toxic or allergenic substances) and phobias (agoraphobia). With such complicated co-symptomatology it is necessary to let the client refer and describe self and so avoid diagnostic criteria. I prefer to avoid the iatrogenic, deleterious effects of putting this label on a person; the curative aim is to take off all such labels. The general rule is the same as for all diagnostic attempts by non-clinical personnel; if in doubt refer the matter to clinical colleagues.

In summary, then, productive depression is a state of calm rehabilitation and recuperation and withdrawal from the hostile world. Non-productive depression is a stuck state with only the beneficial consequences of avoiding the initial trauma.

Stress

I am in a dilemmatous state as I set out to write about the problems and metaphenomena of stress and depression; on the one hand I have to say that I relate to each patient as if I knew nothing about either stress or depression in general – this person is unique and I accept what he or she says about him or her self as he or she fails in self-nurture. On the other hand I want to show that I don't work in a vacuum and that I'm not about to re-invent the subject matter.

At first sight depression and stress reactions are very similar. Each takes some or all of a general pattern of symptoms. Stress usually comes on after a clear cause, while depression may come on many months after a probable cause so that the causal relationship is not immediately apparent and so that there may appear to be no initiating incident. Depression is as much affect as is anger or sadness although of low energy.

No one can live without experiencing some degree of stress ... Crossing a busy intersection ... or even sheer joy are enough to activate the body's stress-mechanisms to some extent. Stress is not even necessarily bad for you: it is also the spice of life. (Selye[16])

There are two ways of looking at stress, the clinical, precisely defined way and the layman's assessment of his or her neighbours. I will pursue the former although I am not going to comprehensively list signs and symptoms. As gleaned from the Selye quotation above, any risk taking during self-nurturing is accompanied by stress. Life without risk would be dull indeed so we all have to learn to cope with stress and the consequences of being unable to do so become the subject of therapy. There is also a threshold for stress; some people seem to be relatively immune so that mild anxiety soon dissipates whereas others seem to simmer, always about to burst into some kind of traumatic response.

A stress disorder is caused either by a major stressful situation or by reiteration of minor situations. Either circumstances may be of obvious negative character, death of a close relative, an earthquake, war, unemployment, failing an examination, or may be generally beneficial character as with a wedding, moving house or changing job. The person's immediate reaction may vary from stoicism to hypochondria. Stress producing predicaments are very common in our time due to the prevalence of non-caring interpersonal relationships in industry, commerce, politics and the family.

The response to stressful circumstances occurs on several levels. Sociologically other people may be involved as helpers or be similarly traumatized. Physiologically general excitement increases <3> and psychologically thought processes race while exploring options to deal with events. One pair of these options is fight or flight as Cannon[17] described; or freeze.

Stress usually occurs in a social situation and the response of the affected person varies according to personal capabilities. Intelligent, vocal people tend to cope <5> better than others.[18] The slow onset form of stress starts with persistent anxiety feelings and thoughts about possible catastrophes and fear of panic attacks which can be dealt with as described on page 175.

Another way of summarizing these events was provided by Selye[19] who proposed existence of a general adaptation syndrome as the reaction to any form of stress. He described the following sequence of events –

- alarm during which the person recognized the stressful state. Adrenaline, cortisol and other endocrine factors are secreted. These result in reactions that are physiological beneficial in reacting to the stress, rises in blood pressure and pulse rate with palpitations, incease in the rate of breathing, increased sweating occurs, eye pupils dilate, the mouth goes dry, some people faint, others can't sleep. Ultimately incontinence of faeces and/or urine occur. As time goes on a headache, backache or other such symptom may develop. The behavioural reaction maybe worsened by panic which is a deleterious familiar figure (page 58).

- **Resistance and adaptation** during which the general state of alertness is maintained but acclimatization occurs and some factors return towards normal, including respiration rate, pulse rate and disappearance of some symptoms. The person may have the impression of being cool, calm and collected, steadily and courageously dealing with the initiating problem(s).
- **Exhaustion** when defences become unable to cope and conspicuous dis-ease follows, including extreme muscular fatigue, heart failure, immunological failure with susceptibility to infection of all kinds and prolonged anxiety and depression. If extended, death occurs.

A paramedic or therapist coping with a stressed person gets the person to a safe place, away from the traumatic event, if that is necessary and possible. This provides the most effective immediate adaptation to the alarm response. The person must be made comfortable and particularly warm. A physician may inject medicaments as he or she thinks fit.

Counselling and psychotherapy, in due course of time, enables the patient to talk about what happened in a traumatic situation and after. Expressions of affect may be very important if the victim considers that other people were responsible for his or her predicament. A confused expression of anger and sadness may be worked through and effective treatment involves reformulating and expanding knowledge of the situation. In the stressed state the expression of alienation may predominate over the expression of centred self and also see Chapter 7.

Post traumatic stress disorder (PTSD) is a major form of stress reaction, distinguished from ordinary stress disorder by occurrence of nightmares and daytime flashbacks in which the traumatic event is effectively relived.

The stressed person is out of kilter, is unable to complete figures adequately and the many books on the subject recommend relaxation exercises (page 175) for restoring equanimity and thus for getting healthy centred figures flowing again.

Tips for coping with stress[20].

- I become used to recognizing stress situations as soon as they develop and deal with them then and there before escalation occurs.
- I know that some things must be endured and taken stoically. E.g., Looking after a sick, cantankerous child or old parent. Set limits to such activities and look after self.

- I relax, the opposite of stress, frequently using a recognized method (e.g., page 175). I relax into relaxation. I avoid *trying* to relax because trying generates stress. I flop out and "go away" to a favourite place in imagination.
- I turn "worries" into "problems" because problems can be solved. I then break the major problems down into minor problems.
- I turn indecision into decision: then act on the decision. Inappropriate action is less stressful than worrying and can be dealt with by a new decision.
- I consider: "What is the worst thing that could happen to me?"
- It is O.K. to express emotion, especially to cry – men, yes!
- I devote as much time as is necessary to keeping well. Avoid medicines such as aspirin and such if possible.
- I promote adventurous interests, hobbies, walking, jogging and other excursions; I have fun!.
- I develop old friendships and make new ones, especially those that involve cooperation with other people. I laugh a lot.
- I am my own best friend. I am good to myself. I congratulate myself when being successful. I accept responsibility for myself and all I do,
- I examine how things are for me and my feelings about them. I deliberately change the feelings from time to time, using music, hobbies, etc. I act into feeling happy – such affect stays with me.
- I share beneficial and deleterious things with a friend or counsellor; I avoid the "Ain't it awful" game.
- I stop drug taking including caffeine in tea and coffee, nicotine from burning tobacco and excess alcohol.
- I read a good book or listen to music I like.
- I spend time recalling happy times.
- I watch comedy on T.V. and laugh freely.

Stress may be the precursor of a further reaction to the initial stress situation, depression, although it is usual for a few months to pass before the new symptoms develop. The more severe the stress the longer it takes the person to get out of the depression.

Hypochondria

The alien games set out above have many things in common. Hypochodria is different. The person concerned is "worried" or "anxious", various words are used, about a real or supposed medical condition. It usually takes persitant chat from a highly knowledgeable person, doctor, etc., to provide reassurance. Two-chair role pay involving chat with the affected part is revealing.

Hypochondria is a constant search for self-symptoms and the fear of illness with real or imagined symptoms leading to passivity, reliance on other people and the doctor. The solution may come from recognizing the symptom in other people.

Passivity accompanied by continuous pathological introspection leads to disablement of the victim.

Phobias

Phobias appear in chapter 7 because the approach using felt-protofigures is very effective and the outcome from the method described there is more immediately beneficial than other procedures.

Hypersensitivity

A person who is wittering around, unable to quieten, unable to be aware of feelings or thoughts, needs to relax.

Meditation to quieten affect. While visualization may entail concentration on a fantasy scene, with or without auditory and/or tactile impression, meditation can be similar or may consist of concentration on something ephemeral, like a mantra. Both methods have the advantage of slowing down racing minds, of reducing the likelihood of affective discharge and allowing free rein for intellectual development, including ideas directed to the limbic system for therapeutic effect. (page 175).

Visualization and meditation are now widely recognized as being very effective in therapy. Thus Simonton and colleagues' wrote—

> Learning to relax and influence the body ... helps people accept their body once again and their ability to work towards health. The body again becomes a resource of pleasure and comfort .... Relaxation helps reduce fear, which can become overwhelming at times with a life-threatening disease.

Since one of the most frequent habits of psychosomatic experimenters is self-discounting, meditation methods to boost self esteem are very valuable.

The body talks

Body states have been well recognized as precursors of diseases and were much studied under the heading of psychosomatic disease. Some examples are set out in Table 5·1. I have extensively reviewed this subject elsewhere[21] so will restrict the following remarks to be highly pertinent to the present study.

A person's life experience determines the person's general character and susceptibility to disease. These experiences include stress situations, both at home and at work. Examples cited[21] included –

- In a large group of women psychological status, as revealed by questionnaires, correlated with ovarian function as assessed by vaginal smears and basal body temperature.
- Young men of apparent healthy disposition fell victim to tuberculosis after unhappy love affairs.
- Young women developed tuberculosis "as the reason why they were jilted".
- A G.P. reported that 75% of diseases he encountered were products of "destructive emotional patterns" and "inadequate mental responses" to them.
- Anxiety attacks accompanying hyperventilation (page 172) were largely relieved by a regimen of breathing adequately: 8 complete respirations per minute.

Table 5·1. Body clues indicating the existence of suppressed figures

Signs or symptoms	Indication
Headaches	Who am I?
Neck and shoulder aches and pains	Where am I?
Knee and leg aches and pains	Where am I going?
Back ache	How do I exist?
Stomach ache	I won't digest (understand) this!
Diarrhoea	Life is running away from me!
Constipation	Don't take this away from me!
Haemorrhoids	I am desperately holding on to what I have got!
Asthma	Pity me!
Clumsiness	Notice me!

- The scapegoat in a family would be the member to develop a psychosomatic illness.
- The person who needed to cry but did not do so was likely to develop a headache.
- Muscular tension, if not dealt with, will develop and become a "cramp".
- Denied pain may reappear when inflicting pain on another person.
- Soldiers who had all been subjected to the same highly stressful battle conditions developed a wide range of clinical conditions, each evidently specific to the man concerned.

The latter observation lead to the idea that each individual had a weak organ or system that would succumb degeneratively to stress situation, correlating asthma and lungs, dermatitis with skin, ulcers with the gut, etc. All rather obvious to us now.

On observing experimenters, the somatic state of the person's gestalt figures is highly indicative of health or lack of health. Thus in health the person is cool, to use modern terminology, while unhealth is restless, ill-at-ease, jumpy and constantly changing or is solid, rigid, unchanging. These dialectically related states may be out of the awareness of the person concerned.

So the prime focus in therapy is active concentration, the opposite to fidgetyness or stiffness. Concentration on the symptom allows the person to discover for self what is going on and then, if he or she will, make a change.

The relief of symptoms is also contingent on resolution of any stress situation and thus of the psychosomatic degenerative condition. Such relief is facilitated by —

- Providing social resources because people who have them adequately experience less distress in difficult situations.
- Decrease the likelihood of the recurrence of the stressful events.
- If the stressing event occurs, other people present

can endeavour to alter the perception of the event in a positive, supportive way and thus mitigate the stress potential.

- Develop coping strategies.
- Enhance personal esteem and feelings of potency in the deleterious situation.

This mini-essay has covered a wide range of psychological and clinical, i.e., physiological, conditions and lends support to the trend of our time to not consider psychosomatic ill-

The impasse

> The principle art of living
> is skill in managing
> interruptions.

Resuming and summarizing the discussion that started on page 30 it can be said that the client/patient (experimenter) is generally unaware of unfinished situations or if he is aware is incapable of coping with them. As emphasized above, retreat to the impasse state of being occurs as a pseudo-resolution of a polarity conflict – usually the spontaneous centred self is overwhelmed by inhibitory, deliberate alienation. This is not always so as two forms of alienation may be in conflict as may occur with F. Perls' top- and under-dog. The impasse is a non-awareness state, is a ground phenomenon, an out of awareness confusion of protofigures. F. Perls' unique discovery was that by patiently staying at the impasse, rather than avoiding it, the alienated self, barren ground phenomenon was subject of a spontaneous transition to the centred state, fertile ground phenomenon. However, Perls' other injunction to "Shit or get off the pot" blocks rather than fosters these processes.

Laura Perls[22] wrote –

> The resistance is something that should not be taken as negative; like in Chinese or Japanese body techniques, you take it as something to go with. Sensing limitations helps you explore the freedom you have within the limitations.

This statement is about wholeness. A resistance is a gap in a gestalt and figure and the gap can only become personal knowledge if fully investigated and not ignored.

The unkiltered person interrupts those of his or her behaviours which are used repeatedly and without discrimination. Hall indicates a source for this in matching, on his diagram, (here as Chart 2·2) each organismic stage with an an-organismic one. His secondary indication is that avoidance at a specific stage has a specific outcome. This is generally not true, as will be discussed extensively later. Thus an unfortunate inference from the layout of Hall's diagram, Chart 2·2, is that continuous progress from one an-organismic stage to another is inferred and this is certainly not true as will be shown by case examples in Chapter 8. The person may become out of kilter at any stage and/or may move to a place that is irrelevant to the initial need. The presenting experimenter may be unable to engage with sensory impres-

sions or, if he can do that, they remain out of awareness. If in awareness, excitement and energy mobilization may not occur or if it does it may be sterile, engagement may not follow. Engagement may occur without action. Action may occur and continue aimlessly beyond satisfaction of need. The need may be satisfied but the awareness of this achievement does not occur, there is no evaluation and no feeling of pleasure. It is rare for such a person to acknowledge the job as done and the return to base may be to the initial sense of unreality, which may itself be denied in a deluge of nonsense, including nicotine and alcohol, inane pop-music, pulp fiction, etc.

Interruptions that became habits were initially useful to the person (L. Perls[23]). When the person interrupts self there may be an abrupt change of verbal subject, a seizure, a stoppage or nothing happens in silence. The person often reports feeling empty, feeling nothing at all. Or, though appearing impassive, reports "mind racing," a plethora of thoughts and impressions too complicated for report. These were the outcomes of healthy processes that were disrupted, interrupted, with avoidance of and resistance to touchy subjects and activities.

This discussion of interruption processes must include disruption by simultaneous, contradictory processes, considered to be the usual generators of the impasse state. While this state is often accompanied by silence during feverish mental processes it is often accompanied by over-generous affect, including laughter or screaming. Expression of this affect is best supported as it burns itself out.

The art in using the figure-stage processes is in wandering about in it. Regression is particularly useful – not a going back in time in the classical way but going back by stages. If **X** has some problems in completing an activity, stage <5>, go back a stage – what happened for him as he engaged with the action situation? Go back further, what happened as his excitement and affect, stage <3> occurred – was this pure anticipation of enjoying stage <5> or did anxiety, boredom, etc., intrude contaminating his potential for enjoyment. Even before that as he internally verbalized his intentions,. etc., what came up accompanying the intention?

In my experience I guided this form of stage by stage regression but did not push. The elucidatory question is: "What happened before that?" – except that it is preferable to specify what "that" is. There may be no outcome in the way presently described. A deleterious circuit may appear – but these will be discussed in Chapter 7. Meanwhile, I will consider what happens at the various stages of figure development when something goes wrong and how to put things right. Other ways through the looking glass.

An experimenter at the impasse is in a state of mind that is rather like that of the experimenter seeking protofigures, as described in Chapter 9.

General ways of dealing with tertiary interruptions: Initiating experiments

Every affection of the mind
that is attended with either pain or pleasure, hope or fear,
is the cause of an agitation
whose influence extends to the heart ....

William Harvey, 1649

Here we have behaviours of the facilitator that must be avoided because they would support the alien self of the experimenter. Other behaviours are valuable because they support the experimenter's core self.

Appreciating deleterious messages

Deleterious processes may have some advantages. Pain is a warning that something must be done.

Presumably sensing psychic pain fulfills a similar function. Acknowledge it! Support centred processes!

supporting centred processes

Perls[7] suggested a series of useful observations that can be made during experiments and be used to guide the experimenter — now somewhat modified. Whenever "over tenseness, cramps, spasms, contractions in your system" are encountered proceed as follows —

- Let the person "Get the proper 'feel'" of what is going on. "Do not attempt any dissolving bef ore 10 to 15 seconds" concentration "on the spot." Watch for the slightest development, like an increase or decrease in tension, numbness or itching. Very promising is the appearance of a slight fluttering or tremor or an 'electric sen sation'. Every change indicates that engage ment is made between conscious and uncon scious instances." Tell the person what you observe.

- Describe the situation to the person in clean language (page 162); suggest amendments to the language of the person to improve clarity.

- Suggest that the person "Take control; relax and tighten-up, by a fraction [!] of an inch, the mus cles in question."

- "Find out the purpose of the contracting. Find out what is being" avoided; get the person to "express" the avoidance. [Perls refers to resist ance and I have substituted avoidance].

- Persevere with what is going on for the person. Other symptoms, avoidances and dissociations may come up. "Make conscious the conflict be tween repressor and repressed. Every discov ery adds energy to the conscious personality: stay with that as much as you can."

- Help the person to "Re-experience the residues of past problems" and "Experience him- or her self – to become aware of all of self, words, ges tures, breathing, etc.

- Concentrate on the "how's", the mechanism of interruption. Avoid "why's".

- Notice and report – does he interrupt his or her sentences, stop breathing, or send energy to unnecessary parts of his body,

- The person can make the key discovery – that he interrupts himself. Is he aware of such self-interruption? Can he discover how self-interruption occurs?

- Celebrate the "ah-ha" experience of discovery, the satori, the emotion of success, when it occurs.

Questions and experiments

The general way for the facilitator to initiate an exper-iment is to ask a question. However questions are, in general, manipulative ways to get another person to do work and counselling and psychotherapy can be, for the experimenter, a grilling with questions. The gestalt approach is not like that since the facilitator relies mostly on process assessments; statements to draw the attention of the experimenter to his or her personal processes. See Appendix 1 for supportive ways of putting clean questions.

Suggestions are tested for reality by experiment. The experiment is validated by the experimenter who observes and reinforces the verbal and dramatic metaphor. These aspects require separate discussion before integration.

When questions are formulated they are best given in the present tense to elucidate ground status and to promote development of beneficial figures (Levitski and Perls[24]). This form of process questioning avoids the pit falls mentioned in the text, including –

- confronting, demanding and manipulating.
- the use of the vague "why" instead of specific questions, what, where, when, how, which. These inquire into the structure of the event.
- the use of "but" and "however" that only gen erate confusion.
- the use of "it", "this", "that", "do", etc., and other words of vague specificity.

Effective questions can include –

- What are you doing?
- What do you feel?
- What do you want?
- What are you avoiding?
- What do you expect?
- What is going on for you, right now?
- What do you want for yourself?
- What do you experience now? (Zinker[25])

- What happened just then?
- What can happen next?
- What happened before that?
- What happened after that?
- What do you gain by ...?
- How do you benefit from that?
- What happened before or after that?
- What would be the opposite of that for you?
- As you quietly meditate on your polarity what happens next for you?
- How do you feel as you do (or say) that?
- As you feel that what do you know?
- What do you know about what is happening for you now?
- I hear you talking about – what do you actually do?
- So that is your outcome – how do you feel?
- So that is your outcome – what do you think about it?
- So that is your outcome — how do you feel and what do you think about it?

The facilitator refines his questions to gain specificity. He aims to increase the experimenter's "response-ability, the ability to choose [among] reactions." The facilitator's primary responsibility is not to let go unchallenged any statement or behaviour which is not representative of the self, which is evidence of the experimenter's lack of self-responsibility. This means that he must deal with each one of the unkiltered mechanisms as they appear. Each one must be integrated by the experimenter and must be transformed into an expression of self so that he can truly discover his core-self. Marcus[26] emphasized that withdrawal of facilitator support was often valuable for the dependant experimenter who is literally forced to find a figure.

As already stressed, the gestalt approach is highly experimental. Is it scientific? Figure and ground observations and generalizations show considerable variations between people and for the same person from time to time. Gestalt observations are reported in discursive form and are not subject of quantitative measurement like weight or height. Just as it took many generations of early medical practitioners to scientifically establish that the fox glove principle, digitalis, was effective in the treatment of heart conditions so it will take many generations of people to unequivocally establish the validity of gestalt, figure ↔ ground, observations.

What if the experimenter asks questions? The facilitator works with both support and frustration and develops this paradox. He or she does not accept questions from emotionally blind experimenters. Such questions can be turned into statements which is likely to lead to self-expression with emotion.

While, in our time, we value both thinking and feeling F. Perls both approved of and discounted thinking processes. At times he criticized thinking by using terms like "elephant shit". He is said to have used this tactic to shake intellectuals out of excessive intellectualization and into engagement with feelings –

> Thinking is rehearsing in fantasy for the role you have to play in society. And when it comes to the moment of performance, and you're not sure

whether your performance will be well received, then you get stage fright" ... given "the name anxiety by psychiatry.

At other times F. Perls encouraged thinking –

> It is at this point that rational thinking has its place; in the assessment of the degree to which catastrophic expectation is mere imagination or exaggeration of real danger. In the safe emergency of the therapeutic situation, the neurotic discovers that the world does not fall to pieces if he gets angry, sexy, joyous, mournful. Nor is the group's support for his self-esteem and appreciation of his achievements towards authenticity and greater liveliness to be underestimated."[27]

Such a double standard is not a paradox, it is the dialectic at work. Perls had differing intentions on different occasions.

Self defence

> Can you feel if the laundry is dry
> when your hands are wet?

> [A]wkwardness and embarrassment are potentially creative states, the temporary lack of balance we experience at the growing edge where we have one foot on familiar and one foot on unfamiliar ground, the boundary experience itself. If we have mobility and allow ourselves to wobble, we can maintain the excitement, ignore and even forget the awkwardness, gain new ground and with it more support. We can see graceful awkwardness in every small child before it becomes socialized and constrained by the civilized demand to "keep it cool." L. Perls.[23]

Relationships between people vary from closeness and harmony, through coolness, to distant, unfriendly, estranged and antagonized dispositions. The processes of change going on can be summarized as separation, withdrawal, disaffection and alienation or the opposite, coming together again, reunification, readjustment and reconciliation.

The problem for a facilitator is to recognise what is going on for the experimenter, while he or she probably does not know him or her self. Processes, here as figure development, are frequently interrupted in healthy everyday life. It is only when a person becomes unable to resume a figure where he or she has left off, that trouble may occur. Though it is not necessarily unalterable trouble. As L. Perls emphasizes above, the wobble occurs during exploration and is part of the learning process. For me tight-rope walking would be a perpetual wobbling process and it need not be so for you.

While the emphasis in this chapter has been on a discussion of tertiary events, foundations will be prepared by considering other, contributory circumstances. These include –
- self-defence as an aspect of self-nurture
- language as a mode of clear or obscure communication with other people
- differentiation of comment and criticism in communication
- the role of fantasy, metaphor and symbols
- the occurrence of obstructive elements including parental injunctions and Dawkin's psychoviruses considered together with curative redecision processes
- the role of memory.

During meditation on these many factors, remember that centred self is fertile and alienated self is barren.

Observing self

I am very impressed by Diekman's[28] arguments for considering the role of an Observing self on the personality map. For me a quarrel between core- and alien-selves can readily be resolved by calling on my observing self to adjudicate and facilitate their processes. Observing self can assess if gestalts are complete or becoming complete and therapeutic use of this stratagem will be further described in Chapter 8.

As an experiment I asked, as my Observing self, what impressions the other two have about all the above. Core-self claimed immediately that he was very happy indeed because he had created it. Alien-self was slow in coming up with some negative comments about language forms and grammar and the naiveté of some the examples (now omited). And anyway he had frequently prompted centred self so he was generally happy with what had happened.

Observing self as self-facilitator

Observing self when functioning, is free from the constraints of gross and ordinary anxiety, is in the calmness characteristic of neocortical cathexis. Clarity thus gained enables options to be explored and self-knowledge exploited curatively. Everybody has ways of clarifying his or her mind including going to someone who will help.

In summary, from Chapter 9, the basic hypothesis is that to free a person from dominance by limbic phenomena draw the attention of the person to neocortical possibilities and facilitate supremacy of the latter over the former.

How shall this be done?

A simple example illustrates although it involves two people: **Vann** was driving through mountainous country along hair pin bends and **F** was his navigator-passenger. By a simple error of judgment **V** damaged the driver's wing mirror of a parked car. After concluding the rigmarole of insurance business with the owner of the damaged car he drove on and **F** could see that **V** was in a highly emotional state. If emotionally preoccupied he would be more likely to misjudge the contours of hair-pin bends. **F** decided to be unemotional and be in an observing state although **F** too felt the trauma of the collision. After a few kilometers **V** drew into a safe place at the side of the road and slept for an hour. On regaining the road he decided to reach the nearest town and **F** could see that **V** was still on the edge of a emotional outburst. So **F** chattered, forcing himself to engage his interest in intellectual things, repeating many of the conversations they had had over recent days. After about ten kilometers **V** parked the car and they negotiated beds for the night in a hotel. Next day **V** was "quite himself again" and they took the road into the mountains.

The point of **F**'s tactic was to help them survive hair pin bends and precipices by ensuring that **V** stayed with his intellectual capacities fully engaged and with his emotional predilections turned off. Only the next day did **F** ask **V** how he felt – the emotional trap – and **V** then shared his anxieties in an excited way

The proposal for treatment of chronic emotional cathexis is, then, that the affected person either spontaneously frees himself of amygdalal dominance or that he be helped to do so by "therapy". Only when emotion is no longer dominant can neocortical processes teach amygdalal processes and this, in **Vann**'s example, was done by himself. The therapist only facilitated the processes. This matter is taken up again in Chapter 9.

Ideolectic indicators

The choice of words indicates where a person is coming from – associated or dissociated self. Expressed another way, language communicates clearly and cleanly with a listener or the listener is confused and may feel a need to ask questions to elucidate what is going on. The effort of exploring this matter is very rewarding.

Pinker[29] while exploring "evolutionary psychology" was concerned with an innate, out of consciousness, proto-laguage the discovery of which he ascribes to Chomsky and calls "universal grammar" and "mentalese." Each child is born with potential proto-language and converts (translates) this into the language of its community largely by imitating its parents and other speakers and then developing and generalizing language proper.

Among the aspects of this new learned language are the traits and mannerisms of the parents, many of them verbal forms that impede rather than promote clear, meaningful communication. F. Perls[24] was evidently an expert in spotting impediments in the people who worked with him and he helped them to clarify communication by word substitutions, using **and** for **but,** etc., as summarized in Table 5·2. This helps inter-person communication and facilitates the persons thinking processes since such ruminations are effectively talking to oneself. The point is to replace ineffective, self-defeating words with effective words – The table also provides clean, centred language transformations from the dirty, alienated language with a summary of what to do about them relative to figure stage formation.

I suggested in previous chapters that you can experiment with evaluating the matter I set down using –

Now I am aware …

and you can do this to assess every factor listed in Table 5·2 and only accept what is true for you. My guess is that you will benefit by cleaning up your language (See Appendix 1). You will become known as someone who communicates clearly and other people will enjoy your conversation.

Censoring. For some people the memory-brain holds an inadequate vocabulary and works slower than the tongue so the system fills in with ums, ahs, and rubbish talk of which swearing is an example of ill expressed affect. Sort out the situation by role play in two positions, 1) let the tongue talk freely and 2) let the ears hear. Swearing has a role in the presentation of emotional states and in relief of tension. Together with the other words expressed swearing is a famil-

Table 5·2. Words indicating the occurrence of interruption of inter-personal processes.

Alienated Word(s)	Effect	Reformulation to centeredness†	figure stage
But *	Negation of a first statement by a second statement: confusion caused by conflicting statements. Remove by using …	**And**	<4>
However *	ditto, with less dramatic effect: use …	**And**	<4>
Try	Stops figure development before activity occurs. Establish "doing" by gentle …	**Experiment**	<4>
Can not	Infers a defect in the action system when the real decision is "will not" = "don't want to." Establish self-support by using …	**Will not**	<5>
Negative *	A positive statement is negated so that two contradictory statements, a double message, is presented. Indicates the presence of a powerful interruption process. Be … Also see Appendix 1, page 171.	**Positive**	<4>
Need	Infers a basic physiological need, drinking, etc. Remove exaggeration by using …	**Want**	<3>
Questions	are often used in a lazy, way to manipulate someone else to do the work. This is revealed if the questioner turns the question into a … Genuine requests for information are OK.	**Statement**	<6>
Why **questions** *	are childishly dependent and avoid personal responsibility by generalizing the questions and calling for a generalized *because* answer. The remedy is to avoid using questions, as above, or to use the specific questions …	**how, what when, where**	<2>
I know	Implies certainty when certainty is questionable and excitement and feelings are avoided. Avoid exaggeration by using …	**I imagine I suspect**	<2>
I feel	Vague affect is expressed when thoughts are intended …	**Express the thoughts**	
It	can be clearly used when very close to the subject it indicates. Otherwise it is not clear what the subject matter is. Communication can be clarified by …	**Specifying the noun**	<2>
Thing **Something** **Someone**	A vague designation is made clear by …	**Naming the subject**	
One	dissociates the speaker from his or her subject matter and from the emotion belonging to the subject matter. Re-establish association and affect by using …	**I**	<2>
You	in the dissociated form, is similar to *one* in effect.	**Use … I**	<2>
We	attempting to manipulate other people into a non-existent consensus. Use …	**I**	<2>
N't	Indistinctly said n't is ambiguous – was it said or not? Use …	**Not**	<4>

* Double messages and generalizations indicate avoidance of decision making, avoidance of self-responsibility, sitting on a fence and leaving the decision making to other people.
† "clean language".

iar figure for many people who may benefit by learning less aggressive forms of expression. The polarity readily changes to 1) the swearer and 2) "I don't like swearing." For the latter suggest a change to a positive formulation.

Generally the form of language used by a person is not under conscious control, it flows naturally. The notable exception is the professional writer who may spend much energy in choosing words and knitting phrases together. So, in the ordinary event, word forms are simple products of memory. Enthusiasm can be hidden by use of weak, vague adjectives like *good*, *bad* and *nice*. The curative move is to draw attention and then suggest reformulation by the person concerned, aiming for elucidation, positivity and specificity. Such language clues indicate the style of interruption used by a person. Others are –

- Use of "pseudo-ego language." *I thought* . . .when there is no sign of thinking; *I mean* . . .when there is no intellectual content; *I feel* . . . when feelings are not evident.
- Use of inappropriate imperative, parental language (I acknowledge that parents need to use this language with children) when dealing with supposedly responsible adults. The words are *should, ought, must, have to* and *got to*.

Body language is often an indicator of the occurrence of interruption processes. Speech impediments are a dramatic demonstration of embarrassment. I can understand a person not looking at me, staring perhaps, but to never look at me – well, am I really there for this person? Does he/she wish I wasn't? Irrelevant limb and body movements occur as blinking, tics, nose and crotch scratching, twitches, fidgety hand and foot movements. Alternatively body movements may be unnaturally inhibited so that the person looks rigid; staring eyes, folded arms, clenched fists and inadequate breathing are examples internally focused and dissociated from companions.

Negative commands. A story, told on page 178, example 4, illustrates the effect of negative commands. Here two figures occur, the first generated by mother's "run into the road" command and the second by her "don't" command. Two contradictory figures are then operating causing the child to be confused and probably to take to ground two ideas, i) that mother gives – always gives? – confusing messages and ii) sends him into danger and is thus not to be trusted. A similar analysis of the events in each of the other examples produces similar ideas about contradictory messages

Noteworthy features of the confusion state are the possible occurrence of high physical energy. The 4 year old in example 2 (page 178) became hysterical and evidently in a state of panic.

The logical explanation for all this is that the negative is meaningless by itself. There must be a positive statement to negate. So the positive statement is processed first and then negated. The important indication here is that the person affected by the negative is not only pushed out of figure, he or she is put into an anxious, confused state from which it is not possible, immediately, to return to figure.

Never voice negatives !
Always voice positivity.

Memory. In these examples the incident is remembered. In Example 2 (page 178) the older boy would be likely to continue, from time to time, to hit the younger boy. The Mother, unaware of the genesis of the initial event, would wonder why her boys were always fighting. She might conclude that that was what all boys were like.

Grievous Mental Harm

This was suffered by inmates of nazi concentration camps and american Guantanamo Bay prison. The consequences of the traumas may appear decades later; see Chapter 9.

- Norwegians who had been forced to work for the Nazis suffered severe psychological stress and developed psychosomatic symptoms 20 to 25 years later.

My defence of myself

When I consider that I am being attacked I have a choice of responding to defend myself or of backing away. My defence is of the self-nurturing abilities of my most real self, my centred self. As a child I began to accumulate defence strategies and these eventually consolidated into a character form, an alienated personality which effectively hid my centred self. My personality is thus an integrand of how I present and express myself and was largely alienated. It could take great daring to expose my centred self.

All will go well for me if my consolidated, alienated, alternative character, compliant or mask-self is roughly in harmony with my centred self or if either one is totally dominant. Trouble occurs if the conflicting interest of this pair flare into open battle. I like to be spontaneous but there are times when I am taken by surprise and I need to think a little, to remember relevant information. If I am asked to do something and I want to be spontaneous I say "No!." It is easy to convert this negative to a "Yes!" at a later time. Much easier than converting a "Yes" into a "No!"

Lost engagements

If someone is feeling maudlin while ruminating over guilt he is blocked off from the vigours and variety of life. He may sense a head ache coming on – so he needs to attend to matters set out in the next chapter where physiological events indicate that gestalts are disrupted.

Notes and references

1   Chekov, A. 1887, Translated and adapted by Hare, 1977, pp. 41 & 61.

2   Mollon, 1996

3   Berne, 1976

4   Perls,F. 1973, p 74

5   Perls, F. 1947, p 223

6   Adler, translation in Ansbacher & Ansbacher[30], 1964.

7   Perls, F. 1970.

8   The red Queen said something apposite to Alice about the meaning of words.

9   Antonunccio, DeNelski & Danton, 1997

10 Gut, 1989

11 A useful general guide for readers who like to avoid psychiatric tomes is Gilbert's essay on *Counselling for Depression.* 1992.

12 For a full review of depression, epidemiology, possible sex and ethnic differences, demographic differences, marital / non-marital differences, employed / non-employed differences, age of onset and the various rates of incidence of the various types, see reference 31.

13 Hare, 1997.

14 Castaneda, 1985, p 148

15 Miller 1956

16 Selye 1956, p vii

17 Cannon 1953

18 Westen, 1998 p 709

19 Selye 1952, 1956, 1982

20 The items in this list have been gathered from many sources. They are set out in first person language so that a stressed person can read them, and, if appropriate, own them and act on them for personal benefit.

21 Edwards. R. (1985)

22 Perls, L. 1994, p 7.

23 Perls, L. 1994, p 91.

24 Levitski & Perls 1970

25 Zinker 1977

26 Marcus 1979

27 Perls, F. 1967, p 14.

28 Diekman 1982

29 Pinker, S. 1994, p 415.

30 Ansbacher and Ansbacher, 1964.

31 Gottlieb and Hamman, 1992, p. 247.

—=(✭)=—

Chapter 6

Physiology and behaviour

When thinking about protofigures
it is as well to remember that
neurology is the brainy
part of physiology.

I really enjoy a full
organismic … sneeze

It is sometimes very obvious that the physiological state of metabolism of a person's body has an effect on the behaviour of the person. As examples, an extremely hungry person or a physically exhausted person sitting panting, trying to get his or her breath back person can't think clearly. People with inborn errors of metabolism can't function in certain physiological ways and also cannot behave like healthy people. The affected person suffers accompanied by the whole gamut of relatives and friends.

At one time this relationship was studied under the heading psychosomatic processes. The first of F. Perls' aims in revising psychoanalysis was "to replace the psychological by the organismic concept."[1] The latter, as is clear whenever Perls refers to organismic processes, is basically physiological. "The laws of support, contact and interruption apply to each level; it is impossible to draw a line between psychosomatic manifestations and psychosomatic illness." Perls[2] emphasized the physiology – psychology connection neatly: "The readiness with which a person considers statements of someone else depends largely upon his oral development and freedom from oral resistances."

Perls[3] found in his practice as a psychoanalyst that Freud's libido theory was inadequate; he found 15% sexual involvement in neurosis and $2 \rightarrow 3$ % in hysteria. Just the same "a neurosis makes sense, it is a disturbance of development and adjustment . . . the outcome of a conflict between organism and environment." Perls[4] quotes Bertrand Russell as drawing attention to the need to analyse the instincts other than sexual, principally hunger. He points out a major mistake in Freud's thinking that the oral and the anal zones are, physiologically, zones of positive and negative hunger and only secondarily sexual. Anatomically the mouth, the place of hunger gratification, is in front, leading to the future. The anus, the place of rubbish elimination, is in the rear, leaving it's stool in the past

While Freud considered the sexual system, Reich the respiratory system and F. Perls the alimentary system the present study includes all physiological systems, in themselves and as metaphors. Starting with the sexual system, we have a clear illustration of one of the aphorisms of Perls; "The neurotic is restless and yet does not change." The sexually satisfied person is quiescent (and may fall asleep) while lack of sexual satisfaction is accompanied by vague physical movement, listlessness, and other features of behaviour that are usually described as "anxiety," as Tennessee Williams depicted in a *Cat on a Hot Tin Roof.*

It is worth reiterating at the outset in this chapter that academic gestalt theory is a theory of perception. Perception, whether mediated by eye, ear, taste, smell or touch, is a physiological process. Anyone who doubts this can refer to a textbook of physiology; Among many treatises I prefer Samson Wright's *Applied Physiology,* a treatise for Medical Students.

A convenient way to examine the psychological consequences of healthy and unhealthy physiological processes is to examine the functions of the various systems, lung function, digestive function, etc. Then can follow discussion of generalizations, including the concepts of the role of equilibrium and homeostasis.

An itch was analysed on page 45 and a similar approach to a sneeze would chart its change from barely perceptible sensation to gigantic ejection of snot droplets.

Physiological functions

Respiration; lung function and anxiety

"Learn to breath properly," wrote F. Perls.[5] The whole rib cage, the diaphragm and the surface muscles of the lower abdomen are involved in economical, effective, breathing. So the person who is breathing properly can be aware of his or her rib cage rising and falling and belly filling out and sinking in. It may be necessary to formally learn to breath and to teach others.[6] Claustrophobia and anxiety are related to breathing problems.[7]

Breathing is biphasic: breathing in brings in and associates with nurturing oxygen and breathing out excretes and dissociates from CO_2. Both are carried by 80% nitrogen.

The connection between the respiratory and gestalt figure stage circuits may be compared by taking the processes of breathing into aware control and considering every step during healthy and unhealthy events.

Organismic, uninterrupted breathing will be considered first, stage by stage of the gestalt figure –

<1> The physiological reality is that breathing continues automatically without need of voluntary intervention, although this may occur. Most breathing is thus continuously at this stage of the cycle; the next, cognitive stage <2> does not usually occur unless something untoward happens.

<2> After exertion I become aware of my breathing and may chose to vary the rate or depth of my breathing.

<3> I can notice that the depth and rate of my breathing automatically responds to my excitement level needs; indeed, my breathing rate is a primary indicator of my excitement level. It is obvious that my mode of breathing responds to my affective state when I sob, shout in anger or laugh.

<4> Engagement; awareness of breathing rate, depth, etc.

<5> Action; knowing that breathing subsists in breathing at the rate and depth determined by my physiological state.

<6> I meet my need and my breathing cycle continues until I die.

<7> I can be aware of my satisfaction as I breath continuously.

<8> Withdrawal effectively occurs at the end of each breath and before the beginning of the next.

<0'> My ground state is satisfactory with respect to breathing when I am in a general kiltered state. The converse is true; one of my indicators of organismicity is breathing appropriate to my energy requirement on the moment.

Temporary breathing stasis may also happen if an emerg-ency occurs. This may be the excitement and surprise resulting in "catching my breath" or intent listening when a strange sound occurs – is it a ghost? – or a tiger prowling nearby? In the latter cases it is of paramount importance to start breathing again immediately or one may not be able to run fast enough.

An-organismic breathing develops from the unnecessary continuation of the needs of those moments of excitement and surprise. The short term consequences are anxiety, panic, sensory blackout The long term consequences of breathing stasis is death. To start with, in stasis –

<1> I stop breathing and produce the oxygen lack and surfeit of carbon dioxide that then produce discomfort and anxiety. The anxious person may not realise that he has stopped breathing.

<2> Under these circumstances I am unaware of my need to breathe.

<3> Excitement may be stifled: I can hold my breath by muscular spasm in my chest wall and diaphragm. If I do that I am not in an organismic breathing cycle. I can discover what else is going on for me in that an-organismic state by continuing to hold my breath.

<4> Engagement with breathing does not occur in stasis,

<5> and no action occurs,

<6> the breathing need is not satisfied,

<7> there is no sense of satisfaction; distress is more likely,

<8> and withdrawal does not occur while the breath is held.

The other an-organismic state is over-breathing which is as above until –

<7> I am unaware of any sense of satisfaction in breathing and so do not withdraw in the sense necessary by slowing down my breathing rate or making my breathing more shallow.

<6> Likewise I am unaware that my need is met

<5> and the action of over breathing continues unnecessarily and I may become dizzy.

A rare form of anorganismic breathing occurs in extreme anxiety states, and consists of the action of breathing in in the chest area and breathing out in the belly area, and vice versa. The net result is no breathing in or out through the mouth.[8] An enormous amount of work may thus be done with no effect. Here the motions of breathing are locked in contradiction so the problem is at stage <5>. Awareness alone, aided by hands on chest and belly, is adequate first aid for such people.

"Breathe!" says Perls.[9] because the person who breathes with his or her whole abdominal and thoracic cavities feels, is contactful, active and happy. Since breathing is essential for life, it is useful to separately consider the psychological consequences of under and over breathing.

Hypoventilation. The physiological consequences of failure to breathe adequately are due to oxygen lack. Grinker's[10] observations are relevant (page 107) and indicate that, if mild, the lack is compensated for; the blood pressure rises, the blood red cell count increases, the heart beat rate increases, the breathing rate tends to increase. The physiological signs preceded psychological apprehension which preceded anxiety and the increased likelihood of thoughts of flight. However, Grinker's students chose to stay with the oxygen-deprivation experiments and avoided escape activity

Chart 6·1. The hyperventilation cycle.

Modified from Cluff[11]

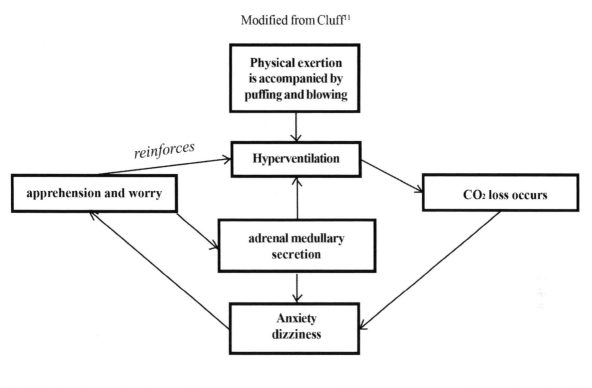

although it was available at any time. If trauma really is a progenitor of stress states and psychosomatic disease, what is likely to be the long term outcome for Grinker's students?

As F. Perls frequently said, the therapist must point out to the client any stoppage of breathing as it is potentially damaging. How is it that the person is doing something that threatens his or her very life support systems?

Hyperventilation. The observations of Cluff[11] (page 107) concerned anxious, depressed or mentally stricken "neurotic patients" with hyperventilation. After being taught relaxation methods and the way of breathing steadily at 8 complete respirations per minute 95% of the patients became symptom free, or relatively so. Cluff's explanation for the occurrence of the hyperventilation symptoms and for the effectiveness of the treatment is based on recognition of a "vicious cycle of self-supporting physiological and psychological components" which he represented diagramatically, reproduced here as Chart 6·1. The brief explanation is that once hyperventilation starts the loss of CO_2 begins to produce symptoms of apprehension and secretion of the adrenal medullary catecholamines, adrenaline and noradrenaline which themselves produce symptoms that are very much like those of the initial hyperventilation, reinforcing the initial symptoms. This self-regenerating cycle of events is reinforced by the effect of CO_2 on haemoglobin which loses efficiency in carrying

oxygen and the resulting hypoxia can lead to dizzyness, fainting and an altered state of consciousness.

Cluff's observations clearly indicate a role for physiological factors in what otherwise can be considered to be a psychological problem. It is to be regretted that Cluff does not give details of the "triggers" that were described to him by the patients as setting off the hyperventilation states. He also does not discuss the role of psychological factors, care and personal attention, as provided by himself and his colleagues who taught the patients to breathe adequately.

Perls described a patient (Appendix, Example 3, page 108) for whom teaching of adequate breathing and relaxation relieved the symptoms of asthma and high blood pressure.[12]

Eating; oral and alimentary function

Hunger is the "feeling" aspect of the assimilation of food. As F. Perls[13] recounts, assimilation in the gut occurs most readily after effective biting, both incising and masticating. The end point of assimilation coincides with the end of the feeling hungry. (page 109)

The natural manifestation of destructiveness, aggressive tendencies, in Perls words, biting food is expressed and gratified two or three times each day. The person who swallows food whole, however, will not obtain adequate gratification and will tend to project it out onto the world or tolerate it and suffer the uncomfortable lump in the belly.

Just as the anorexic person is without feelings of hunger, so the lethargic person has lost interest in the world. The adipotic person cannot satiate hunger and the over lusty person cannot satiate sexual contact with other people.

Perls[14] made much of oral function and its disabilities including discussing paranoia, mental anaesthesia, as the psychological analogue of oral anaesthesia. While attention to flavour and texture, biting, chewing and masticating are important the process may start with proper use of the knife and fork on the plate.

Junk food, fast food, is one of the problems of our times and the computer operators rule can be modified to work here –

oral garbage in → oral garbage out

both on the physical level and psychological level and as metaphor. People who eat junk food regurgitate junk-verbiage. Disgust is an oral function. I spit out foul food; I refuse foul information. An unhealthy person may swallow foul food, vomit some of it, and wait patiently for diarrhoea to supervene. Foul mental input leads to mental vomiting and diarrhoea. If vomiting is incomplete, retching may continue for life. Perls[15] analyses disgust in the way depicted in Chart 6·2, neatly summed up, I think, as: "don't be disgusting in front of us. Nice people don't retch, belch, fart, vomit and shit in public." True of course and a heavy load for a small child to carry.

"Correct eating … is of vital importance in achieving an intelligent, harmonious personality," in removing the "Bottle neck of mental inhibitions." And disbelief and avoidance of this idea is, in itself, indication of the activity of "dental inhibitions and deep seated neurotic attitudes." The therapeutic move is to attend to growth of suckling and biting into full mastication. Perls' sequence[16] is –

- "We have to be fully aware of the fact that we are eating." That means only eating, without reading, chatting, day-dreaming, worrying, etc. The eater can observe tastes and textures without comment and judgement.
- Bite large pieces into small pieces using incisor teeth. Continue to observe taste and texture.
- Chew small pieces into a pulp with masticator teeth. Continue to observe taste and texture.
- Swallow only when a fluid mush has been attained. At this stage the food is being drunk.

This is the gestalt circuit of eating. By proceeding to chew, with full awareness, to fluidity interest in flavour and texture is maintained and boredom avoided. If, in the experiment, concentration on taste and texture wavers, let it. Stay with the fantasy, etc., and then return to taste and texture. Alternate attention with the intention of extending the duration of savouring of the food to the full time of eating. Thus is introjection avoided.

An attack on someone by biting can be exercised in fantasy; ensure both biting and chewing. There will be a decrease in the number of fears and phobias. Foul food must be vomited (Chart 6·2; Perls[17]). The alternative is a life of verbal retching. Therapy involves staying with what occurs, retch, vomit, belch, fart, shit, piss, get rid of all the putrid stuff. Cry, scream, be angry, be afraid, get rid of rotten garbage.

Toxic substances in the diet. have been recognized to have profound influences on behaviour (Sluckin,[18]; Edwards[19]).

Karl presented with mild anxiety, "depression." and/or continuing or intermittent tiredness, "heaviness" and lack of energy to use in activity. **Karl** needed to investigate, in group, the way he prevented mobilization of his energy. An alternative approach that is becoming increasingly recognized is that there may be toxic substances in the diet. These substances vary from those of natural origin, such as wheat germ in coeliac disease, to synthesized chemicals that are added to foods for cosmetic reasons or to "increase shelf life." The effect of these substances varies from clinical allergy to the mildly debilitating symptoms of food intolerance[20] described at the beginning of this paragraph.

It may be necessary to attend to longer term dietary effects. To use my experience as an example, I found the following list of substances to be toxic for me; caffeine, some varieties of honey, all citrus marmalades, the colours azorubin (E122), sunset yellow (E110), tartrazine (E102).

The sensory signal from the rectum indicating that it is time to defecate is often suppressed because it comes at an inconvenient time. This signal is known as a pain, particularly in childhood,

An interesting avoidance process occurs in nice society, as it has immediately above. If the words "faeces" and "defecate" are used, the pungency of words such as "turds" and "shit" are avoided, the knowledge of disgust is also avoided.

Muscular control: Stuttering

Among many events when muscular control is confused or lost, stuttering (or stammering) is most common. It occurs occasionally in anyone who becomes excited without a clear sense of what needs to be achieved.

Pinker[21] pointed out that the construction of complex sentences requires holding grammatical forms in memory until the moment of use occurs. The limit here is again Miller's 7±2 units of remembered whatever[22]. Anyone with a score much less than 7 is going to fall over themselves if trying to expound complicated ideas. The therapists recommendation can be to slow down and simplify the ideas to be expressed.

A profound example of less than 7±2 confusion is the stutterer, who, in my experience, is trying to express two (at least) ideas at the same time. **Finn**, a postgraduate research student came for counselling when he had a lecture to give before a learned society. He had two months in which to learn to express himself clearly and without stuttering and stammering. In 1:1 relationship with **F** the latter encouraged him to slow down and express his competing ideas one after the other and completely separately, giving each enough time. An example was talking about his anxiety about the up coming lecture while also being concerned about what he would have for lunch and where.

After five weekly sessions he had gained in articulateness and eventually gave his lecture perfectly; his head of department who heard him did not know about the therapy and was amazed. This example illustrates gestalt figure analysis at work. While it is possible to express two figures through two separate channels, voice and foot kicking were an example of F. Perls', it is not possible to clearly articulate two events simultaneously through one channel, as with the stutterer. The solution, as illustrated above, was to articulate the figures separately. We have interesting contrast of technique here because F. Perls[23] treated a stutterer by mock strangulation – a more existential approach.

In the more generalized situation errors made by hands or feet – carpentry, piano playing, dancing – may be due to influence of more than one gestalt in the communication channel. The solution to the problem is presumably similar to that above – slow down the activity, and express each of the gestalts separately and in turn.

Mind/Brain: mental confusion

Referring to the stuttering example it is clear that much muddled behaviour can occur under the influence on the mind of more than one gestalt, in fact of multiple protofigures. The solution to this problem is to slow down, meditate and concentrate on one idea only. If there is more than one idea, keep them distinct and separate from one another.

Some techniques may be helpful for getting into the meditation state but ultimately meditation would be concerned with the problem of the time. My mind tends to wander so I periodically bring my attention back to my problem.

Sexuality; genital and endocrine function

The views of Reich were summarized on page 34 and in Chart 2·1 and were, in 1960, one of the pregenitors of this essay. Genital malfunction in sexual abnormality (so called, impotence) responds to psychotherapy. "The man with premature ejaculation has weak boundaries and is not in contact with his partner. He is inclined to blame his partner because he projects his activity onto her."[24] Manifestations of stage fright, shyness and shame are analogous to failures of penile erection.[25] Little boys are often punished for showing off their "willies." Failure of vaginal secretions and other female events occur in similar fashion. According to Reich it is beneficial to be freely sexual.

The observations of Benedek (page 176) can be seen in chicken and egg form, which precedes which, which is causative to the other, does endocrine function determine psychological state or *vice versa*. By seeing a psychosomatic unit, Benedek found explanation in a unitary, holistic phenomenon, much as Grinker had with his students. Benedek did not discuss wider social factors, such as the reaction of

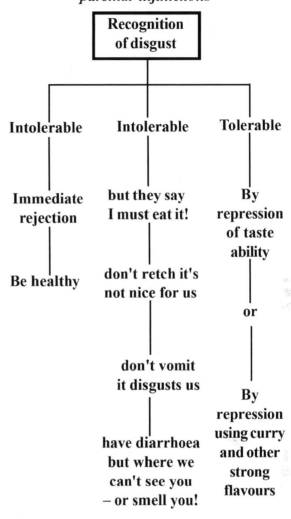

Chart 6·2. *Dynamics of disgust; the role of parental injunctions*

husbands or other companions, the families and other events such as pre-menstrual tension.

Gus came to 1:1 therapy complaining that he had not had a penile erection since his lover died of Aids. Over three visits dialogue with his penis cleared the matter for him, enabling erections with a new partner and provided a new dawn for him.

Relaxation; muscle function

F. Perls wrote ambiguously about the value of relaxation. In 1947 he said[26] that relaxation is insufficient because the body armour returns on return of excitement. In 1966 he said[27] that relaxation during therapy is beneficial and will help repressed material to come to the surface. Perhaps he meant that relaxation by itself was of limited value and that it is only beneficial in the context of on-going therapy.

Relaxation can become a task that has to be determinedly carried out and then becomes ineffective; rather relax into relaxation. This was my experience with cancer patients who tended to try very hard at first at anything that caught their interest.

Insomnia; sleep function

Insomnia is a healthy, organismic reaction to abuse of living status, so said Perls.[28] A healthy night of sleep occurs if the business of the day is finished before going to bed – or if it is at least put in abeyance until tomorrow. Saying good bye on leaving the business premises is beneficial. Insomnia is part of the agitation expressed on being locked into fixed figure states; in old terminology, insomnia is neurotic behaviour.

The occurrence of micro-sleeps during the day is also a part of the abnormality patterns of anxiety states, though mental exhaustion, as during a boring but interesting lecture, is another causative factor.

Immunological failure

Day (page 108) observed connections between the onset of tuberculosis and the wish of a group of men to avoid the responsibility of earning a living and a group of women to avoid the consequences of jilting. The social need became manifest and then disappeared in retreat to a mountain top to suffer what was then a fashionable disease and to receive continuous nurturing from others.

With hind sight one can comment on the failure of the physiological, immunological system in the context of psychological and social stress. Perls' report of the woman with bleeding gums is somewhat similar. Here the avoidance was focused in the direction of her husband and her mother.

Hypochondria and hysteria

This man presented himself as having hypochondria explaining that he could pick up, emulate and manifest any combination of symptoms. Thus his physiological component was variable. His psychological component probably was based on simple identification with his father.

Trauma and stress; general debilitation

Perls and colleagues[29] discuss trauma and also call it a "bad habit" and a "complex." They discuss the following aspects (my breakdown and indication of gestalt figure stages) –

• there was something desired – the need <1> and <2>
• There was a danger sensed in the satisfaction <7> of the expected desire or need
• The desire or need was frustrated
• The tension of the frustration became unbearable
• The desire or need was deliberately inhibited
• and so was awareness <2> of the desire / need <1>

Table 6·1. *Other physiological functions*

System	Defect	Effect on behaviour
Pancreatic	Insulin deficiency	Comatose unless given insulin. Mild form is hypo- or hyperactive.
Renal	stones	very painful; needs surgery
Colon	diahoea	moribund needs anti infection treatment
	constipation	lethargy; needs laxative
Adipose	excessive fat	diabetes and heart disease needs strict diet
Anorexia	lack of food	moribund
Hepatic	blocked bile duct	very painful; needs surgery
Thyroid	Lack of T4 and T3	Lethargic

• Thus was suffering and danger avoided.
• The "whole complex" of feeling and sensory impressions <3>, expression <4> gestures were unfinished.
• Said "complex" is now not available for use.
• Considerable energy is expended in keeping the complex out of use – keeping it in its fixed, unsatisfiable state.

Noteworthily absent from Perls' analysis is mention of engagement <4> and action is only mentioned in the form of gesture <5>. Are Perls and colleagues implying that trauma always and only occurs at stages <2> and <3>?

Other physiological conditions are summarise in Table 6·1. With each the degree of debilitation, and thus effect on psychological state, varies from minor debilitation to lethal. As emphasized above, if in doubt suggest that the experimenter consults his or her general practitioner. The example discussed was the investigation of felt-senses in the stomach area when it may be necessary to be sure that the stomach tissues are not ulcerated.

Interim summary

Physiological dynamism has both direct and a metaphorical bearing on the psychological state. The gestalt figure factor is often not immediately clear but, if taken on trust, can provide the basis of beneficial therapeutic effects.

Contemplation of the behaviours associated with closely related psychological states illustrates the close link of psychological and physiological states –

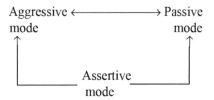

On a quiet summer night
A far off woman's voice squawks
"Life is a Cabaret".
An incongruent, destructive metaphor
to rock all possible boats.

Generalized physiological functions

Cycling is O.K.
What do I do
When I get a puncture?

Anderoo

Physiological control by homeostasis

The organism has psychological *contact* needs as well as physiological ones ... these ... are met through ... the psychological counterpart of the homeostatic process. (Perls[30] emphasis).

[T]he homeostatic process is the process of self-regulation, [is] the process by which the organism interacts with its environment. (Perls[31]).

The principle activities of the life maintaining processes of both animals and plants are assimilation and excretion. Plants assimilate water and minerals from the soil and CO_2 from the air, aided by sun light energy, and excrete oxygen. Animals assimilate food from plants or other animals, water and oxygen and excrete waste as sweat, piss and shit and CO_2. The quantitative balance of carbon and oxygen in the atmosphere between plants and animals consist of –

Plants

CO_2 O_2

Animals

This is the cyclic metabolism of Lovelock's Gaia. It is also the process being thrown out of equilibrium by industrial and other CO_2 overproduction leading to global warming.

Assimilation and excretion are thus the polar, dialectic processes supporting life. Death occurs if either fail.

The basic kiltered control processes of the body at the level of physiology and biochemistry maintain particular and general states of equilibrium and health. Running produces a great increase of heart rate which returns to normal, to its equilibrium rate, on resting. There are also concomitant and similar rising and declining changes of blood glucose turnover, respiration rate, endocrine function, etc. This is the *homeostatic* process described by Barnard.[32] and Cannon.[33] Perls[34] emphasized the basic importance of this idea and restated Bernard's proposition as follows –

"... all life and all behaviour are governed by . . homeostasis, .. which the layman calls adaptation."

Homeostasis is the "process by which the organism maintains its equilibrium and therefor its health

under varying conditions . . . and the process by which the organism satisfies its needs. Since the needs are many and each need upsets the equilibrium the homeostatic process goes on all the time. All life is characterized by this continuing play of balance and imbalance in the organism. When the homeostatic process fails to some degree, when the organism remains in a state of disequilibrium for too long a time and is unable to satisfy its needs, it is sick. When the homeostatic process fails, the organism dies."

Equilibrium

Homeostasis is movement about an equilibrium state. The level of glucose in blood circulation, in health, stays relatively constant, at an equilibrium level. If muscular activity takes glucose out of the blood stream, the glucose level falls. In health the original state is soon retrieved by glucose production in the liver and elsewhere. Homeostatic processes thus include movement away from equilibrium and movement back again.

Healthy functioning is thus supported by a myriad of equilibrium and homeostatic process involving, as already mentioned, water, salt, oxygen and CO_2 and also calcium, phosphate and many other dietary substances.

F. Perls makes much of homeostasis and the maintenance of equilibrium states. But I think he misunderstood the physiological facts when using them as a metaphor and I think it important to become clear on this issue. In health the concentration of salt in our blood plasma is constant within very fine limits – nearly the same as that of sea water. This is maintained by hormonal control. A person who takes a lot of salt in his diet passes the excess into his urine so that the blood level of the salt remains constant. The homeostatic factor here is the secretion of salt excretion stimulating hormones from the adrenal glands. In adrenal gland failure homeostasis fails and a new false equilibrium with lower salt concentration occurs and the person becomes very ill and will die if not given synthetic replacement hormone.

It is clear that as behavioural processes homeostasis and equilibrium are intimately related. The members of a family adjust their behaviour, each to the others, so that fairly smooth functioning occurs. Homeostasis entails father not provoking mother too much and vice versa. The children also do as they are told within reasonable limits of behaviour. A natural equilibrium occurs, mother cooks, father mows the lawn, etc. with occasional swings away from equilibrium which are followed by homeostatic return. If father takes to drink and become aggressive all other members of the family adjust, finding new, false equilibria with heightened tension and antagonism – unless mother suddenly says: "Enough is enough!" and walks out.

Boundaries in the physiological field

While the whole body is the physiological field the most important areas are those concerned with meeting the environment, the external field, where the life supporting activities occur, including the air spaces of the lungs and the wet contents of the alimentary canal as external functions, and including elimination of waste and the reproduction of the species. In addition, as said above, to being life supporting these systems are metaphors for the psychological systems.

As emphasized above, Perls[35] gave much attention to alimentary tract function starting with suckling and biting and going through to excretion and adding sexuality. One of his aims was that the "biological function of aggression" should be established as "the solution of the aggression problem," I find fault with this terminology since lung, alimentary and other functions are essentially assertive and not aggressive.

Physiological figures pertain to primary needs, hunger, thirst, etc. and each of these may have sub-forms. Quenching hunger in the infant is suckling which Perls[36] like Freud, relates to morality and ethics. Available breast is pleasure and absent breast is pain. Availability of gratification links directly with sensitivity to trauma and "badness." Both ends of the alimentary tract are involved in tactile and erotic sensory input. The sexual organs, essential for the survival of the species, have function on internal and external boundaries. The testis secrete testosterone into the blood and pass spermatozoa through the prostate and penis to the outside world. The ovaries also secrete hormones into the blood and pass ovae through the fallopian system into the vagina which is effectively external world. The skin, which excretes sweat, is also an erogenous area of variable sensitivity and thus internally an adjunct to the nervous system. The pituitary, adrenal and thyroid glands are internally functional only but are essential contributors to a zippy, healthy life style.

As is well established in physiology and biochemistry, the mechanism of the changes of homeostasis involve cycles of stages of activity and some cycles may have of the order of twenty stages. An exemplary cycle is the adrenal cortical control pathway of glucose metabolism. It is part of the means of maintaining the blood glucose level within close limits, part of glucose homeostasis. Another aspect involves insulin secretion by the pancreas.

Cascade is the term physiologists give to an amplifying system, for example, where quantities rise stage by stage, from micrograms of ACTH from the pituitary gland in the base of the brain to hundreds of grams of glucose in circula-

tion. In terms of energy the figure stage cycle (Frontispiece) is a cascade from stage <1> to stage <5>, very low energy felt-protofigure to knocking a ball for six.

The unhealthy physiological field

There is little benefit in offering psychotherapy to a person who is ill due to water, salt, glucose, etc., loss. These must be attended to first.

Extending the argument beyond hunger and somewhat randomly for brevity, death may be –
- due to oxygen lack occurs on a scale of a few minutes
- due to water lack in days
- due to failure to excrete excess water in days
- food lack in a few weeks
- failure to excrete faeces in many months

I was taught, as written above, as a trainee student counsellor, that the first move in counselling was to ensure that the student had enough to eat and drink and a comfortable place to sleep. Injury from chronic failure to feed adequately, as in a famine, produces the clinically recognized illnesses called anorexia, marasmus and kwashiorkor.

Perls[37] ascribes to Ferenci, supported by Reich, the idea that "the closing muscles of the anus is the manometer of resistance." These "are brought into play in order to avoid the feelings of disgust, etc., part of the avoidance of emotional expression." Haemorrhoids are common in our time.

Physiological metaphors

While the early physiologists observed cause and effect relationships, such as between electrical stimulation and muscle contraction and pancreatic insulin secretion and blood glucose levels, the modern tendency has been to recognize holistic phenomena in the form of long chains of interwoven relationships. An example is the pathway for control of blood glucose level involving the hypothalamus, the anterior pituitary gland, the adrenal cortices and the liver. As mentioned above, such a cycle of events is truly organismic since any departure from equilibrium during activity is followed by a return to equilibrium when resting, an example of homeostasis as described above.

There are four aspects of physiological cycles that are particularly interesting.
- Energy relationships are such that each step involves a small energy increment, which sum to the equivalent of a large increment.
- Most steps of the cycle involve positive stimulation; thus cortisol stimulates hepatic gluconeogenesis. At least one stage is inhibitory; here higher glucose levels in the blood have a negative

effect on the hypothalamus. It is the balance of the positive and negative effects, spanned over space and time, which tend to keep the whole process in equilibrium.

• The circuit functions as a whole entity, irrespective of the number of stages and, indeed, the whole differs from any possible sum of the parts; no stage in glucose metabolism controls the turnover of the cycle, the whole cycle does that.

• The overall effect of a cycle can be amplificatory in terms of quantities. Thus, the quantity of substance involved at each stage of the cascade, as it is called by physiologists, is about ten to one. The hundreds of grams of glucose utilized in marathon running are liberated as a result of processes involving a quantitative amplification factor of the order of 10,000,000.

This is a clear example of the dialectic at work. Feedback is a negation process, abundance of glucose leads to decrease of hypothalamic peptide secretion and, eventually, to diminished glucose secretion. On the other hand, if glucose is removed from circulation, for example by increased muscular work, inhibition of the hypothalamic peptide secretion eases off and renewed glucose secretion occurs.

This dialectical aspect can now be generalized –

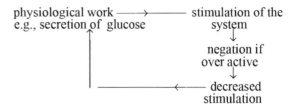

While one can see stage by stage cause and effect relationships the overall effect is holistic. Lack of health results from a break in a physiological pathway. Thus a very small number of children are born each year with an inborn inability to biosynthesize aldosterone – a condition called congenital adrenal hyperplasia by the histopathologists. An affected child will die within hours if not treated with an analogue sodium retaining hormone. The physiological state has its sociological manifestation if the person is ill; the lives of parents and other relatives and friends are disrupted.

So –

Failure of breathing, eating and sexual expression interfere with general behaviour. All "neurotic" conditions interfere with breathing, eating and sexual expression. Psychology and physiology are one within the human frame.

The interaction of physiological processes and psychological processes

Example 1: Events produced by oxygen-lack

This matter was discussed on page 100.

The physiology of oxygen-lack was well described by Grinker[12] and was based on observations on students as experimental subjects. Reaction occurred in several stages –

1   localized reaction
2   generalized reaction consisting of mobilization of haematopoetic system and increase of red cells in circulation.
3   vascular dilation and increased heart pumping rate
4   as CO_2 accumulates respiration rate increased
5   psychological changes occurred, apprehension was followed by anxiety
6   talk about moving to somewhere where there is more oxygen; the possibility of flight.
7   It may be presumed, says Grinker, that ultimately a state of panic and shock will occur in which the limbs are flailed around randomly with considerable violence.

Thus the endeavour to keep a steady internal state of oxygen lack lead to an escalating series of organismic ameliorative steps The stage of psychological interest is that of "free anxiety" in which neither fight nor flight occurs and indecision operated as paralysis. If it was possible to effectively attack, and thus avoid, the extreme symptoms of oxygen-lack, this did not occur.

A form of psychological avoidance occurred. In psychoanalytic parlance (Grinker) this was flight into psychoneurosis. Grinker compared these observations with others he had made in a war situation when a catatonic stupor supervened.

Example 2: Events produced by oxygen-surfeit

This matter was discussed on page 100.

Cluff[2] described the situation of approximately 1,500 patients with chronic hyperventilation. While the patients were, from a clinical diagnostic point of view, a heterogeneous group and "generally classified as neurotic, i.e., with anxiety states, anxiety depression, ... panic disorders, etc., in which complete remission, whether spontaneous or under treatment is uncommon ..." "The common feature is a background of physical and psychological symptoms which can be shown to be associated with hyper-ventilation."

It was found empirically that patients, irrespective of diagnostic category, could be taught not to hyperventilate and thereafter lost their somatic and psychological symptoms in the vast majority of cases ... 70% were completely symptom free, 25% had only minor symptoms, and only 5% failed to respond. (Cluff[2]).

Cluff gives most attention to physiological and psychological mechanisms and questioned the patients about "Trigger" mechanisms and about known ways of stopping the hyperventilation.

The treatment is described by Cluff as physiotherapy and consists of teaching relaxation (Jacobson's method[14] page 175) and after awareness of breathing patterns and rate when hyperventilating, of teaching a slow rate of breathing, 8 complete breaths per minute, with use of hands to check that all parts of the chest and diaphragm are involved. These breathing exercises were to be carried out twice a day for 20 to 30 minutes and were continued until the pattern was established as habitual or for up to six months, whichever event came first.

Example 3. Speechlessness

This example, discussed on page 101, shows two contrasting physiological components, immunological failure in asthma and neurological failure in frigidity and high blood pressure.

A lady continued with me after her former analysts had discontinued treatment because of negativistic and aggressive attitude. She had originally started the analysis because of the high blood pressure, a chronic pseudo-asthma, frigidity, and family difficulties. So great were her breathing difficulties when she started with me that she could scarcely speak. First, I decided to tackle her asthma and postpone work on the deeper personality disorder. After a few hours of reorganizing her breathing, she burst into tears of deepest despair, and with this she obtained her first relief. Three months later her asthma and high blood pressure had disappeared, and now after six months she has lost her frigidity. At present we are working on her self-consciousness. One experiment in particular brought home to her the mechanism of her armour. At a distance of about ten feet from me she was relatively at ease: upon coming nearer, she stiffened more and more and again lost tenseness with distance. This reaction worked in an entirely automatic way. It was necessary to make her realize that visualizing the approach of somebody produced the same effect, and further, that she was not only stiffening but that she was also stifling something. (Perls[15]).

Perls does not say whether she was hypo- or hyperventilating but makes clear that he thought that eliminating the breathing difficulty was enough to resolve all the behavioural and other difficulties.

Example 4. Psychosexual factors

This presentation is referred to on page 103.

Benedek[14] followed the psychosexual lives of a large group of women. She used psychoanalysis to assess psychological status at daily intervals. An endocrinologist independently assessed ovarian function by taking daily vaginal smears and the basal body temperature.

A year later a third person conducted the correlation of these observations. The correlation provided a high degree of predictability of hormone status from psychological state, or *vice versa*, throughout the ovarian sexual cycle.

Benedek concluded that the manifestations of the sexual cycle constituted a psychosomatic unit. Later observations extended the menstrual cycle observations and showed that the psychosexual – ovarian function correlation was also valid throughout pregnancy.

Example 5. Life-style stresses

This presentation is referred to on page 104.

The following patients showed the power of social factors.

A dozen years ago Dr Day began to puzzle over the problem of why so many apparently healthy young men and women were arriving at his [tuberculosis] sanatorium. They were just the age when their physical powers should have been at their peak for resisting the bacillus. They had not been exposed to the conditions which are supposed to facilitate the progress of the disease.

Dr Day rejected the thesis that it might be pure chance that one was struck down while others escaped. He began to ask questions and he found that a surprising [to him] number of cases were patients who had been involved in unhappy love affairs. He looked further, and saw some other emotional difficulties coming at a time of life when young people are meeting their first problems of independence. The emotional stresses involved are quite severe.

Dr Day noticed that girls who had been jilted . . . could make themselves and others believe that they had to give up the young men because of their illness. The unhappy youths could escape from their entanglements or the necessity of earning a living because of the sad state of their health. The women who yearned for attention and luxury found a rather poor substitute, but still a substitute, in the care which tuberculosis brought them both in sanatorium and at home. (Dunbar[15]).

Dr Day is thus reported to have made epidemiological and sociological observations with psychological superstructure concerning the onset and progress of a bacterial infection.

A kiltered gestalt figure is a unique unit
different, distinct from any other figure.

An aberrant figure shares characteristics
with other aberrant figures; anxiety and worry.

Notes and references

1   Perls, F. 1947, p. 82.
2   ibid, p. 125.
3   ibid, p. 14.
4   ibid, p. 94.
5   Perls, F. 1947, p. 266.
6   Victims of the British military system (in my time anyway) were taught to hold the lower belly in a rigid, out of breath state. This was said to stop development of feelings, thought to be necessary preparation for battle. These people need special help as they do not realise how disabled they are.
7   Perls, F. 1947, p. 267.
8   Edwards, 1994.
9   Perls, F. 1947, pp. 266 → 267.
10   Grinker, 1973.
11   Cluff, 1984, pp. 855 → 862.
12   Stammering and stuttering involve gross interruptions of breathing; and see page 102.
13   Perls, F. 1947, p. 110.
14   ibid, p. 164.
15   ibid, p. 199.
16   ibid, p. 194.
17   ibid, p. 198.
18   Sluckin, 1993.
19   Edwards, 1996.
20   Brostof, 2008.
21   Pinker, 1994, p 201.
22   Miller, 1956
23   Perls, F. 1975, p 168
24   Reich, 1960, p. 167.
25   ibid. p. 177.
26   Perls, F. 1966, p. 226.
27   Perls, F. 1966a, p. 230.
28   Perls, F. 1947, pp. 259 → 261.
29   Perls, F. 1951, p. 344.
30   ibid, p. 7.
31   Grove and Panzer, 1989. Grove's method is akin to Erickson's[38] approach to hypnosis in which no formal induction of trance state occurs.
32   Barnard, 1949.
33   Cannon, 1964.
34   Perls, F. 1947.
35   Perls, F. 1949, p. 108 & pp. 113 →118; 1947, p. 192.
36   Perls, F. 1947, pp. 53 → 56 & p. 58.
37   ibid, p. 74.
38   Erickson, Rossi and Rossi, 1976.

—=(☆)=—

Part 3

Innovations: Gestalt Psychodynamic Analysis

Many current challenges to human health and well-being can be
accounted for by the inconsistencies
between environments in which we live today
and those in which human
physical, emotional and psychological needs evolved.

Travathan, W.R. 1999, p 183, *Evolutionary Medicine*, Oxford.

These final Chapters will be devoted to reviewing important
developments of practice for mending the holes in gestalts by
considering and treating –

Grounding of ungrounded sequences and circuits of gestalt
figure stage-like events, such as occur in depression, in Chapter 7.

Evolutionary aspects and influences on nurture, reproduction, mutual cooperation, pleasure and continuation of our species
in chapter 8.

Traumas and similar problems by employing the development
of protofigures to form new gestalts in Chapter 9.

Ruminative commentary on the substance of this book in
Chapter 10.

I see, hear and feel.
I interpret what I see, hear and feel.
I make sense of what I see, hear and feel.
I change what I see, hear and feel.

I can only lead
in my direction
of travelling.

Chapter 7

Ungrounded Circuits

The purpose of poetry
is to mystify
in order to demystify
        W. H. Auden

It was in 1986 that I first noticed that an experimenter, taking his time, was speaking a monologue about his misfortunes in a repetitive way, cycling back to earlier, similar and sometimes practically identical statements. Thus alerted I began to observe repetitions in the processes of other experimenters who, to some degree, were not adequately nurturing themselves. Each, if left to continue, would go round a cycle several times. As I became bored by such repetitions I drew attention to my feelings and observations. If the experimenter denied the validity of my observations, even if fellow group members supported me, I played back the audio-tape recording. (I always recorded group sessions).

After much more experience I realized that these experimenters repeated whole chunks of information in the same sequence of parts without variation. To make matters easier for myself, and to aid the awareness of the experimenter and other group members, I began to map out these emotional manifestations and intellectual observations as they occurred on a flip chart.

Feed back from the experimenter was always the same, something like: "Yes that's what I am always doing though I hadn't realised it in detail before." And sometimes, additionally, something like: "It really gets very tedious. I not only do this kind of thing during the day but go over it time and time again when I am trying to go to sleep at night." It seemed to me that this was, for the experimenter, a very valuable increment of awareness.

As events occurred in a group context other group members, about one third of them, began to be aware of cycling during their presentation of themselves. So a disadvantage must be admitted here; the ongoing group process became one of learning about cycling processes, a biased situation in which presentations by some group members were no longer spontaneous. Justification for continuation in this way was provided by the personal benefit accruing to the experimenters. Further bias was provided by having a large replica of the gestalt figure-stage Chart (Frontispiece) hanging on the wall.

I propose now to share with you my observations on the nature of some of these circuits. My account starts with the index experimenter, mentioned above, and then I present others in roughly an ascending order of complexity, not in order of presentation. While 26 experimenters are discussed here an approximately equal number are not available for presentation due to lack of notes written at the time, of audio-tapes of sessions or of flip-chart evidence; some experimenters took the latter home with them for further study or contemplation.

There is sufficient variety without repetition from experimenter to experimenter to justify anecdotal presentation of all 26 in order to impress on the reader the importance of the observation of cycling activities. Most frequently left out of this account are details of the standard Perlsian transformations as described in Chapters 4 and 5, including owning what is said by using *I* instead *of you,* speaking in the present tense, avoiding "but", employing positivity, being specific rather than using "it," etc.

Method

In the Charts of the following experimenter presentations the aspect presented first is depicted at the top of the chart and lightly shaded. As explained in the notes on the index case, statements and events during presentation were then and are now reduced to little more then slogans on the charts for brevity. The facilitator usually, but not always, selected the words and the experimenter approved or modified them.

As stressed in the Introduction (page 6) the identity of the experimenters has been disguised. However, each is given a *nom de plume* for use in the text and the initiating letter is used in the dialogue report.

The story of each experimenter is here related in much simplified form. Many experimenters presented cycles of stages twice or more though this is not emphasized in the account. Each took several hours of close attention, sometimes spread over two to five weekly group meetings. With the few exceptions detailed, no attempt was made to take a formal history since this would bias the experi-menters interest towards that of the facilitator. Only what the experimenter presented spontaneously was regarded as relevant.

A degree of artificial dispassionate attitude has been introduced in the accounts in this chapter by the author referring to himself as **F**. As facilitator of the processes of investigating cyclic activity **F** found it useful to ask one of two basic questions: "What happens next?" (Grove[1] and page 149) or "What happens before that?" These questions were presented in the present tense with the possibility of bringing the experimenter's experience into that present time. The questions were also, with benefit, sometimes combined in trance inducing form (Erickson and colleagues[2]): "What happens before or after that?"[1]

Index experimenter; Jan, Chart 7·1

This man, in his mid-twenties, came to group complaining that he did not know anybody and did not know how to get to know anyone. On his third attendance he was gazing at the floor reminiscing and it was with him that I noticed the repetition, in the form of a complaint about having no family or friends, followed by a statement about depression and then his misery. His dread of a useless, doleful, lonely future was followed by a frenzied declaration about his panic attacks which sometimes occurred as he was going to bed. Panic did not last long and was followed by lack of energy which he called "being flat," and then the return to the moan about being without a friend and abandoned. I decided at the time to let it pass and a few minutes later the repetition happened again so I decided to share my observation and feeling with him. "I notice that you are repeating yourself and I find it boring." "That's how it is" he said. He wanted to leave things as they were stating that he had had more than his share of group time.

Before the next group meeting **F** set the stages out on a flip Chart, here as Chart 7·1, using the audio-tape to check accuracy. The following week **Jan** looked at his Chart and said that he remembered the stages but that some of it was no longer true.

F   As you think about the stages of your six stage circuit of factors I wonder if you expect that one of them would be more easy to change than the others?

J   [Read the stages one by one and on the second time round stopped at "nobody notices me"] People do notice me. Too much sometimes. I wish they would mind their own business.

F   Does anyone here notice you more than any one else?

J   Yes, indeed, **Netta** does [a group member near him]. I wish she wouldn't.

F   Talk directly to **Netta**.

J   And you do, [looking at **F**] but that's your business.

F   Slow down a bit. What are you most interested in, **Netta** herself, **Netta** 's intrusiveness or me?

Jan then spoke directly to **Netta** but not in terms of objecting to her noticing **Jan**. Rather the opposite as he shared appreciation for her concern. **F** was fascinated because **Jan** was no longer isolated and antagonistic but was smiling and looking at **Netta**. After a few minutes of the **Jan ↔ Netta** interaction, **F** suggested that **Jan** engage with other group members which he did with some warmth.

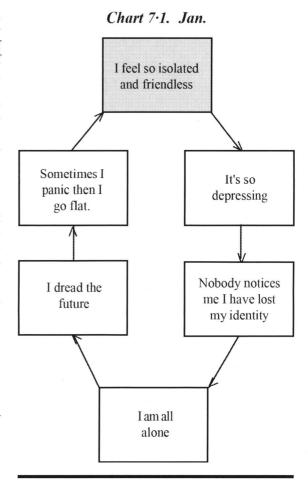

Chart 7·1. Jan.

The following week he went round the circuit of stages on the chart again while talking about recent events –

J   I'm still a bit depressed. The people I like notice me and I like that because they seem to see me as somebody with an identity, with character that can be liked. I have felt sad sometimes, these past few days, but certainly not miserable. It's as if having friends isolates me from such misery. The future seems OK for me though I have a lot to sort out. There was just one night I felt panic before I went to sleep and that was when I had an interview with my boss the next day and thought I would be unable to present myself properly. But I fell asleep OK. And I was OK with my boss.

F helped **Jan** to go over some of these points in detail, sorting out options for further activities and deciding what to do if feeling sad or beginning to panic.

J   I'll phone my friend **Kelly**. He is OK to listen to my woes and I listen to his.

Jan stayed in group for three more weeks, going over some of the aspects of his circuit factors until he felt positive and clear about them.

Referring again to Chart 7·1 I want to emphasize that brevity when making the chart and now as I set out this account have lead to distortions of what was really going on for **Jan**. "It's so depressing" was said with a falling voice

Chart 7·2. Leo

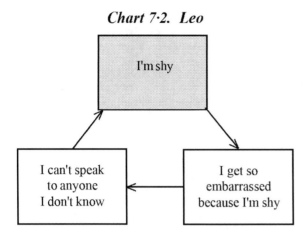

Chart 7·3. Leroy

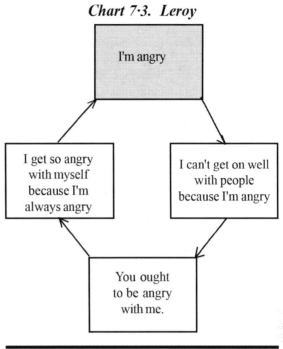

tone. "Nobody notices me" is short for a long statement about going to work, etc., and being at work where only his immediate boss spoke to him and then only about work. Being "miserable" was accompanied by a rather theatrical wringing of hands. "Dread of the future" was accompanied by gazing at me as if to say, so I surmised, "Save me, please!."

As said above, **F** constructed the chart from the information on audio tape and could have separated "desperate" and "miserable" into two boxes. Likewise "panic" and "go flat." **F** had made two discoveries. One was to become aware of the recycling and the other was to spontaneously suggest that the chain of stages had a weak link. **F** became excited and pushed **Jan** into recognizing the vulnerable stage. In this sense **F**'s role in the analysis was not dispassionate and indeed **F** was intensely i nterested and involved. In his excitement **F** was somewhat judgmental.

Leo, Chart 7·2

This late-teen man presented with concern about his shyness which involved bright blushing. It was some weeks before he spoke in the weekly group and then promptly produce his physiological symptom. He then dealt with his bullying father, who demanded both silence and speaking up. He mentioned depression among his concerns. After some weeks of sporadic working he produced the three stage circuit as depicted. By this time he was more forthright in group and with little blushing. On challenge by **F Leo** was able to blush deeply at will which lead to an ability, so he reported, to diminish his blushing in public.

F   Which of these three events is the weakest, where might you best make the changes you want?

L   I want to speak happily to anyone and everyone.

Over a year he practised this in group, particularly with new members and began to report success with people in his work place and pub. He cut himself off from his father so that the bullying was not reinforced. He met his mother at his sister's house. He continued to blush, though not so brilliantly. His hesitant speech became more fluid though he developed a pronounced stutter which was dealt with by another procedure. (page 102).

Leroy, Chart 7·3

This middle aged man was very gentle and polite on arriving in the group. He would sigh and talk weakly about his bad feelings and anger with most events and people in his life. When directed at other people, focused on a cushion, he shouted loudly and jumped about while thumping vigorously.

F   What is going on for you?

L   I feel uncomfortably angry [pause] with you – you got me into ridiculous theatricals. I did it all badly – aren't you angry with me?

F   [Sensing that **Leroy** needed to de-confuse himself by expressing ideas about his events]. No, not angry; a little sad. How does your anger benefit you?

L   There you go again, manipulating me. [After a pause] I'm confused about anger with you and anger with myself and your lack of involvement with me. It's very depressing.

He then produced his four stage circuit as he began to focus his anger on himself.

F   Which of these four factors do you want to deal with first?

L I don't want to be angry with myself. There are plenty of other people who can be angry with me.

He role-played some of these people and readily turned to humour with each one, saying "how ridiculously ineffective such anger was." After 3 weeks he reported that he was now aware that he got on better with people and that he was not aware of the old "suppurating undercurrent of anger." Except, that is, with certain people with whom he had "genuine resentments." These were explored in the usual way (page 85). He also checked several times that **F** and

Chart 7·4. Lord

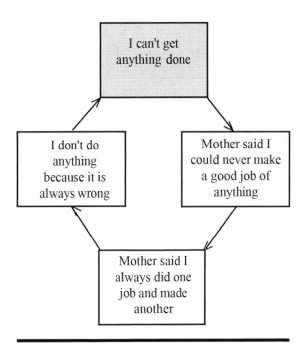

other group members were not, and had never been, angry with him.

Pearl said she had been frightened when **Leroy** was being angry in the group. **Leroy** seemed genuinely very sorry about that and at **F**'s suggestion role played **Pearl** in dialogue with himself.

Lord, Chart 7·4

This man in his early 60's had talked in group of his depression and the week before working on his Chart had produced a dream which, simplified, was –

L I am on a fire-escape and want to go from it into the building to which it is attached at the third or fourth floor level. But this fire-escape doesn't connect to the building. I can see another one connected to the building and need to jump across – I am terrified as I wake up

F How can you safely get from one fire-escape to the other?

L I know these fire-escapes. They were attached to different buildings I worked in. [He described them in some detail]. I was scared as I moved from one job to the other.

F I want to repeat my leading question – how can you safely get from one fire-escape to the other?

L I don't want to jump. I would fall and hurt myself.

F Perhaps the fall would not be so disastrous, this is a dream.

L [After a short pause] I could go down the first fire-escape, walk across, and climb the other one.

F So?

L I am doing that. [Walked round the room]. I'm O.K.

and in the building.

F Where something completely different will occur. Do you feel finished for now?

L Yes, indeed. Very interesting.

When a week later he produced his circuit the stages were not directly concerned with depression. He had been hesitant in saying anything in group and produced his circuit in emulation of **Ned** (Chart 7·14). He was then very expansive with details of things that went wrong for him.

F So you never get anything done, you blame your mother although you obey her injunction to "do one job and make another." Where on your Chart would you expect to make an impression and change the flow of events.

L I wish mother had held her tongue.

F That's all very well. What would you have her say?

L You make mistakes sometimes but I like it when you get things right.

F Which of these two factors would you like to elaborate on, making mistakes or getting things right?

L Oh the getting things right business. That's what I want more than anything.

The process continued over three weeks, referring to Chart 7·4 each time with **Lord** reporting on experiments he had set up at home (his mother was not there) and at work.

F also set up experiments in group in co-operation with other group members in which **Lord** carried out successful, creative experiments.

His hesitancy in group fitted his cyclic presentation as anything he said or did was likely to be wrong. He talked himself into making good jobs of anything he did. It was interesting that his dream illustrated his hesitancy and inability to make a sensible decision. Working through the dream did not give him satisfaction – the circuit work did and he became bolder.

Nona, Chart 7·5

This middle aged woman came to group on the recommendation of a previous member and had seen the Chart this person made while working on depression. Having told us about this she complained immediately of depression.

F As your depression goes on what happens for you next?

N I can't remember anything at all – it's terrible. I wander round to get on and do things and forget what I want by the time I get there.

N debated this theme at length and it was the following week when she spoke about being fed up about being unable to remember and this lead on round her circuit to depression again. The next week, on contemplating the Chart, she expressed amazement that she hadn't realised that her feelings of depression were connected with her poor memory.

F There are several stages on your Chart, remembering what you want to remember, getting depressed when you don't want to, so that life does not work for you. Which do you consider would be a suitable stage to give attention to, memory, making yourself fed up or making things work for you.

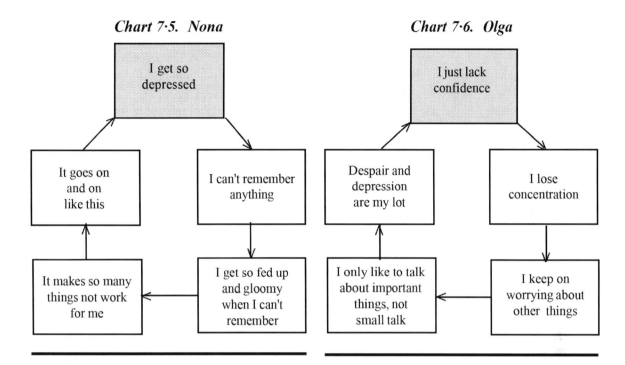

Chart 7·5. Nona

- I get so depressed
- It goes on and on like this
- I can't remember anything
- It makes so many things not work for me
- I get so fed up and gloomy when I can't remember

Chart 7·6. Olga

- I just lack confidence
- Despair and depression are my lot
- I lose concentration
- I only like to talk about important things, not small talk
- I keep on worrying about other things

N [After some hesitation]. Memory, definitely, how can anything else work if I can't remember?

F One thing that you certainly remember is that you can't remember.

At this point **Rab** (Chart 7·21) cut in –

R I had memory trouble and someone told me what to do. I did two things, one was to have a little note book and write things down, just a word does usually, and two, I keep saying this word or something like it, over and over as I go to do things. Works a treat. Why don't you try it?

N That seems a good idea.

F I add to **Rab**'s superb suggestions the act of thanking your memory every time it works for you.

In the group **F** set up experiments for **Nona** to remember things while other group members distracted and interrupted her. (**F** did not remember to suggest the procedure detailed on page 173). Over time she became more successful, had many good laughs and did not again discuss depression.

One great advantage of group work is that everyone has the possibility of producing wisdom.

Olga, Chart 7·6

This woman in her thirties came to group complaining about lack of confidence. When talking about her experiences she wandered from subject to subject with little sign of completing anything, stressing her worries, lack of optimism and concern about anything other than her church affairs. She found help, she said, in singing hymns and praying but

many good feelings then generated soon evaporated. **F** then prompted her into investigating her circuit.

F When you find yourself thinking that you are not confident, what usually happens next?

O I don't like being like that, I lose concentration, my mind wanders off and I worry about all sorts of things that I don't really need to be concerned about. So I decide to concentrate on something important and when I talk to people I hate small talk. People don't like that and take no notice of me and I feel so depressed. It's really all lack of confidence that is to blame.

Olga, having spoken of confidence went on quickly to fill in a lot more detail on a Chart about her concerns with concentration and worry. Then, at the beginning of another group session –

F Look at your Chart. You have made this over the last couple of weeks. Is it all real for you?

O It's just something else to worry about. If I do all that all the time what will become of me.
Yes, it is real enough for me.

F Do you want to change all this?

O Yes. I despair of ever feeling OK.

F What do you want to deal with first, with concentration, your worrying, your dislike of small talk, your despair or your depression?

Olga went on extensively on the misery of all this. It was not until the following week that she said she had been thinking about her family and how they reacted when she would only talk about the important things from the television news bulletins. She had begun to talk about "silly stuff" like the weather and had found pleasure in the other person's

Chart 7·7. Pearl

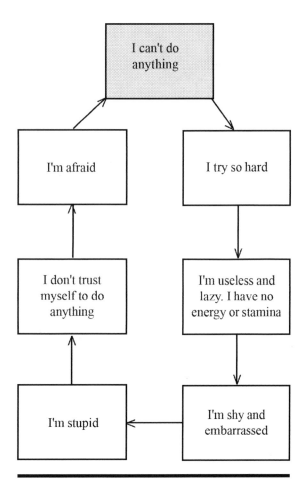

fellow interest.

In group she explored light conversation with group members giving, on **F**'s suggestion, special attention to looking eye to eye. Finding it "stupid" to talk to a young man about his rather garish shirt, both got into peals of laughter.

F noticed that the social aspect seemed to be the affective factor as **Olga** experimented with connection. She "lit up" as another person showed genuine interest in her, irrespective of subject matter. After a few weeks of experiment in group she found a man in the group who discussed both Wittgenstein and cricket and she had a good laugh at the "ridiculousness of it all." She looked at her Chart and reported that it "seemed odd" that she was once without confidence and concentration.

Pearl, Chart 7·7

This woman in her mid-twenties appeared to be transfixed by what Berne called the "try hard" injunction with the result that she was very inefficient at doing anything.

Having completed a seven stage circuit, she focused on trust as being most interesting and most likely to become changed.

P   I can trust myself. I can, will and do trust myself in any situation. I'm feeling quite bold.

At **F**'s suggestion she turned to various group members and told them how she could trust herself to do various things. At one stage she walked around from person to person, adding that she was not afraid.

F   What is the opposite of being afraid for you? I suggest that you make a statement starting: "I trust myself and …."

P   I am full of courage and I can just get on with what I want to do without pushing myself.

Over three further weeks in group she reported on her various successes in being relaxed while getting things done in the outside world (page 175).

F, having been T.A. trained, was at first eager to concentrate on the "try hard" stage. Fortunately he kept this to himself.

Lou, Chart 7·8

This middle aged man expressed himself very philosophically until he began self-deprecation.

Rex [Looking at **Lou**'s Chart]. I relate to your bossiness. I don't like it when you come on strongly to me about something. You say you are ignorant and useless and so am I but there is no reason for you to pretend that you are worse than anyone else.

L   So what must I do, then?

F   You can do something now. Want some suggestions?

Then followed the familiar scene in which the words "should, ought, must, have to and got to" became challenged in **Lou**'s vocabulary using instead forms of the verb "to choose." It took some weeks for him to gain proficiency in avoiding the "bad words" and he noted as time went on how he was enjoying life without thoughts about depression and being over burdened

Another group member chose **Lou**'s weak link for him. **F** would not have intervened so didactically.

Poppy, Chart 7·9

This young woman evidently spent most of her time being confused and unable to decide what she wanted to do for herself.

F   Are you mostly concerned about your confused state?

P   Well I suppose so. Though I would like to stop that anticipation game I play. I am continually trying to forestall criticism and this means that I spend a lot of time thinking about what people want me to do and then doing it, although that is not usually interesting for me

F   What would you be doing if you were not anticipating criticism?

P   I would do what I really want to do.

F   [After listening to **Poppy** going into detail about what she would like for herself.] I have an experiment to suggest to you. Would you like to hear about it?

P   Yes, go on with it.

F   Put someone who criticises you on this other chair,

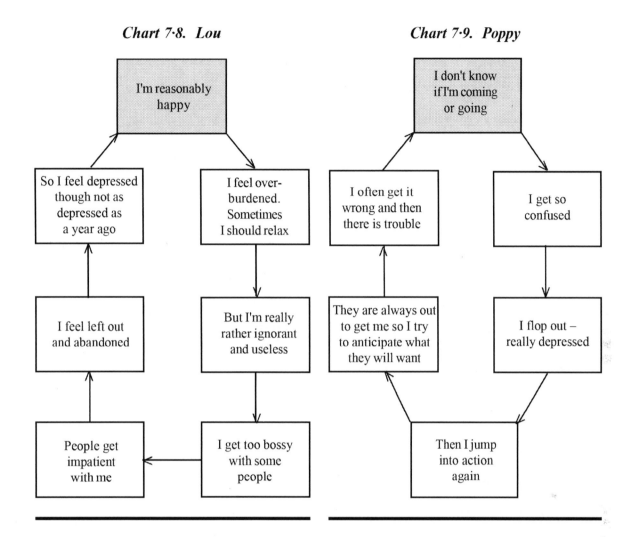

Chart 7·8. Lou

Chart 7·9. Poppy

tell us who it is and tell them what you want.

P   Mother, I don't care if you do criticise me. [To **F**]. It takes much courage to challenge her. When I was a child it was essential that I avoid trouble from her as she beat me. It is not so necessary in adult life.

F   Tell your Mother that.

P   I don't need to avoid trouble with you as I'm more likely to beat you than you beat me. [She laughed and continued with a much detail of how she would deal with criticism from Mother, Father and an older sister.

F offered to share his view about the relationship of comment and criticism (page 87) which **Poppy** heard and appeared to find very interesting.

F   Are you expecting criticism from me?

P   No, you are too gentle.

F   From anyone else?

P   Yes, Lucas looks at me with strong disapproval

Then followed a dialogue with **Lucas** in which they, at **F**'s suggestion, reversed roles for a time.

F   Are you anticipating anything else, here and now?

P   [Indignantly] No, should I?

F   Look at your Chart. You could, on the old way of things, be expecting something and perhaps doing something to anticipate the results.

P   I don't understand you. I'm not into that anticipation thing just now.

Poppy came to group four more weeks but didn't focus on anything in particular. She reported more success in relationships at work although at home things were awkward as they "don't seem to have noticed that I have changed my attitude."

Marty, Chart 7·10

This young woman was diffident about taking part in the group before detailing the large number of domestic items she wanted to buy and which would make life easier for her.

F   How do you propose to get what you want?

M   I don't know. I could do various things but nothing works out. I went to the market to look for second hand things but they were all rubbish and the stall holders were so pushy. They don't want to help. I have no money, no job, and I desperately want these things, 'fridge, microwave, new tele' and camcorder.

F   As you look at the details of your Chart does it occur to you that you might tackle one factor more easily than another.

Chart 7·10. Marty

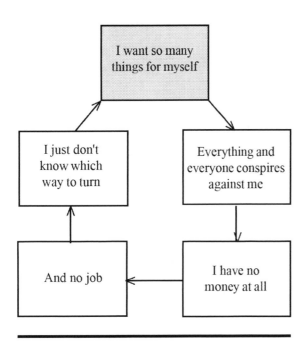

M [Continued to detail wanted objects, then –] It would be easier if I had a job and an income though I do have some idea of which way to turn. It is no use looking for a job, there are no jobs for a woman of my little experience. I could focus carefully on one objective; I used to do that years ago, I could save a little of my money so that I could go to the coast and find a summer job. [She discussed details of where and when].

F Does anything conspire against you realising the objective of working by the sea?

M No, my folk won't like me to go too far away but I know that I will enjoy the time off part of working in a holiday place. I only have to sell ice cream, or something during the day and will be free in the evenings.

Marty missed many group sessions and came again after Easter. She said she was going to Eastbourne for the summer where she had a cousin and said that life was easier since she was concentrating on one thing at a time.

Max, Chart 7·11

This young man had been caught taking goods from a supermarket without paying for them. A friend had suggested therapy might help him change his habit of stealing. When **Max** first spoke in group he went straight into detail about his stealing and he seemed to be proud of his skills. On his second time of talking about his being scared of being caught, **F** began to construct the Chart. Then, having stopped further repetition –

F I want to recap. You steal as an adventure and for excitement and then feel low until you steal again.

M That's how it goes. Stealing can be very exciting, I'm really keyed up and happy. Then I get maudlin so I have to hype myself up for new go.

F If you were to give attention to the stages on the Chart you have constructed, which stage do you think would be likely to give so that you would change your behaviour.

M You think I want to give up thieving. Well sometimes I do and sometimes I don't. It's enjoyable and better then working for a living.

F I have an experiment to suggest for you. Would you like to hear about it?

M O.K. but don't expect me to do as I'm told because I've had enough of that from the Police, Magistrates and Probation Officers.

F I suggest that you set up a dialogue between you and someone you stole from.

M No way. That's just silly stuff.

F Well, set up a dialogue between the aspect of you that thieves and the aspect that has trouble with the thieving.

M O.K.

He spent time over several weekly groups without clearly resolving the thieving ↔ anti-thieving or scared of being caught ↔ not scared polarities. After a month or so another group member reported hearing that **Max** was in prison, presumably for the offence that had occurred before he joined the group.

Roma, Chart 7·12

This young woman quarrelled readily with fellow group members who tended to avoid her.

F I notice that you seem to have trouble relating to other group members. There was quite a noise during tea break.

R He [pointing at John] was telling me how lucky I was to be here with you. I don't like it. I'm fed up with him and everybody I know.

F What are they all saying to you?

And then began the Chart construction. The next week –

F As you look at the details of your Chart does it occur to you that you might tackle one factor more easily than another.

R Everybody treats me like a child and I don't like that. [Turned to John]. Don't talk to me as if I was a child!

F Repeat that statement getting louder and louder.

Roma did so and on **F**'s suggestion said it also to other people present including **F**. Later she role played John and gave herself a dose of put down words. The following week she reported –

R I've been saying to myself: "I'm an adult, a fully grown adult." It's odd, people seem to hear me although I don't say it out loud.

F Who suggested this to you?

R Nobody. I did it.

Chart 7·11. Max

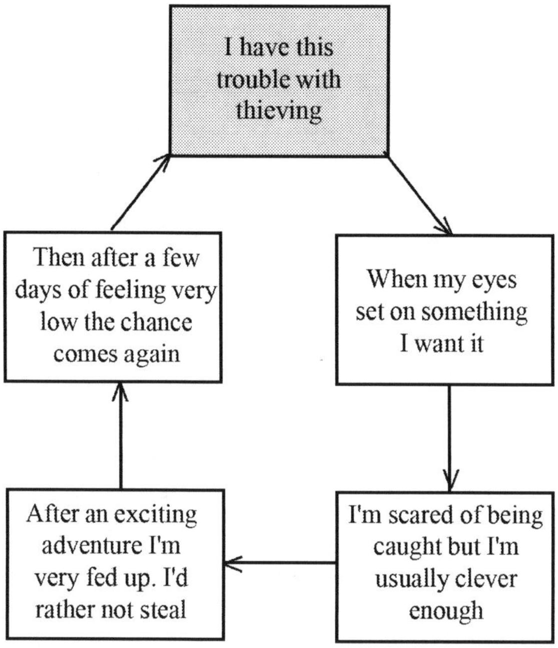

```
        ┌─────────────┐
        │ I have this │
        │ trouble with│
        │   thieving  │
        └─────────────┘
         ↗            ↘
┌──────────────┐   ┌──────────────┐
│ Then after a │   │ When my eyes │
│ few days of  │   │ set on       │
│ feeling very │   │ something    │
│ low the      │   │ I want it    │
│ chance comes │   │              │
│ again        │   │              │
└──────────────┘   └──────────────┘
       ↑                  ↓
┌──────────────┐   ┌──────────────┐
│ After an     │ ← │ I'm scared of│
│ exciting     │   │ being caught │
│ adventure I'm│   │ but I'm      │
│ very fed up. │   │ usually      │
│ I'd rather   │   │ clever       │
│ not steal    │   │ enough       │
└──────────────┘   └──────────────┘
```

Chart 7·12. Roma

```
        ┌─────────────┐
        │ I'm fed up  │
        │ with the    │
        │ people I    │
        │ know        │
        └─────────────┘
         ↗            ↘
┌──────────────┐   ┌──────────────┐
│ I don't like │   │ They         │
│ being treated│   │ continually  │
│ like a child │   │ tell me how  │
│              │   │ grateful I   │
│              │   │ should be for│
│              │   │ the help they│
│              │   │ give me      │
└──────────────┘   └──────────────┘
         ↖            ↙
        ┌─────────────┐
        │ They        │
        │ interfere   │
        │ more than   │
        │ help        │
        └─────────────┘
```

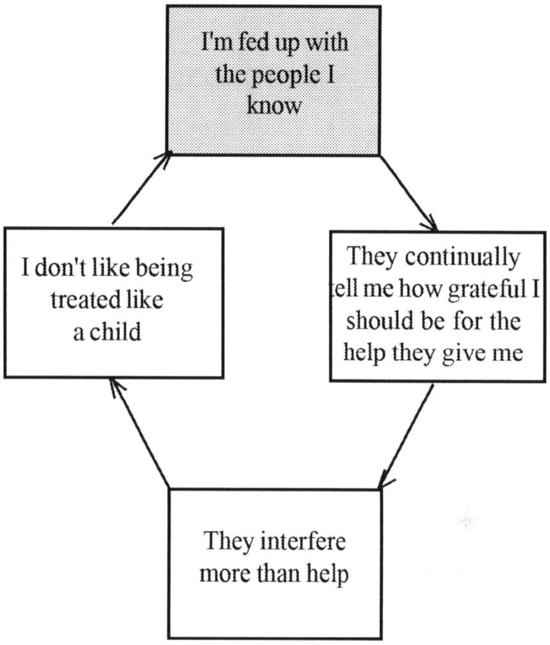

Roma produced her own positive invocation and used it Coué style. It is possible that the idea was generated in the group by herself or another person but **F** likes to think that she generated it as she said at home or at work, thus expressing in words something about her disposition that induced other people to treat her as an adult.

Neal, Chart 7·13

This young man in his early twenties took some weeks to be happy enough in group to participate. He seemed to be waiting for something. His "Life is just a dress rehearsal" statement was thrown off flippantly.

F   What happens for you before you say that?

N   Oh, I don't know! Yes, you don't know what will happen next and it doesn't seem to matter as I am really getting ready for something else.

F   So what happens next?

N   Nothing much. I could tidy my room, or something. I just wander around looking for something to do and don't see anything interesting. This annoys my mother who annoys me. Very depressing and I've no idea what will happen to me.

Neal chose to concentrate on "I don't know what will happen next." He quickly changed it to "You've only got one life." On **F**'s suggestion he changed this to "I've only got one life." to which **Neal** added "And I'd better get on with something."

F   [**Neal** had been shouting at himself] You could treat yourself gently.

N   I really do know, all the time, what I want to have happen next. I seem to want to spite my mother and do nothing, just to show her.

F   Show her what?

N   I am boss. I will decide what to do.

Then followed the familiar battle with mother on cushions and the outcome was "Life is the real play, on stage all the time with a real, live audience. Playmates here I come!"

F was wary at first with **Neal**'s sloganizing and concluded that this was his way of expressing himself. Perhaps he was a budding actor.

Nudd, 7·14

This young man was clearly scared as he joined the group. Some weeks later he confessed that he had come to the door for two weeks running before he actually entered.

When his work did begin he hopped quickly from scared, through paralysis and panic to feeling useless. None of these was directly connected to being scared and it was two weeks later that –

F   What happens before you feel scared?

N   I worry about doing something wrong. As time goes by I could make a really bad mistake and ruin

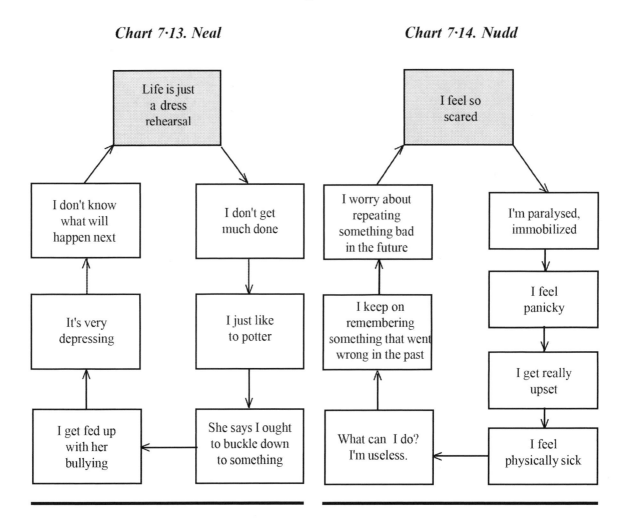

Chart 7·13. Neal

Life is just a dress rehearsal

I don't know what will happen next

I don't get much done

It's very depressing

I just like to potter

I get fed up with her bullying

She says I ought to buckle down to something

Chart 7·14. Nudd

I feel so scared

I worry about repeating something bad in the future

I'm paralysed, immobilized

I keep on remembering something that went wrong in the past

I feel panicky

What can I do? I'm useless.

I get really upset

I feel physically sick

everything. This is because I keep remembering how useless I am.

The chain completed he went round it for some time giving examples of how things went wrong for him.

F   So everything always goes wrong for you!

N   I do remember things that went right, of course things are never always wrong. And I know that when things go right people like me.

He elaborated with many examples, becoming bolder and bolder.

The following week he went back to the Chart and pointed to "I'm useless" saying "this was true sometimes but I'm really very capable." He gave some examples and then said that he was "not feeling so upset these days." [Referring to his Chart] I never really panicked or was paralysed; it was just a passing phase, I suppose.

Rona, Chart 7·15

This young woman followed the example of another woman and quickly constructed her chart, spending most time on despair and being "right at the bottom" and "in the pits." which turned out to be a variety of depression.

She literally vigorously attacked the messages on the Chart

making facile modifications as she proceeded. "Regrets" became "I'm pleased with so many things," giving many examples and proclaiming that she had chosen "New paths" without specifying what they were. She gave attention to being at the bottom saying it was not true since she was, again, on "New Paths." She crossed off "depressed" in red pen and wrote "live" alongside it. **F** enquired and was told that this word was the imperative of the verb "to live."

The following week she told us about her wonderful week end "with this monstrous rugger player."

F did not believe that anything significant was happening as **Rona** rushed headlong and head strong, throwing ideas in all directions. He also did not believe in the reality of an obliging rugby football player. However, at the end she showed no signs of the misery, etc., she had talked about. Perhaps her powerful fantasy was an incisive metaphor.

Otto, 7·16

This young man produced two circuits on successive weeks, marked **P** and **Q** on the Chart. Like **Lou,** trust was an important issue for **Otto**: they did not know one another.

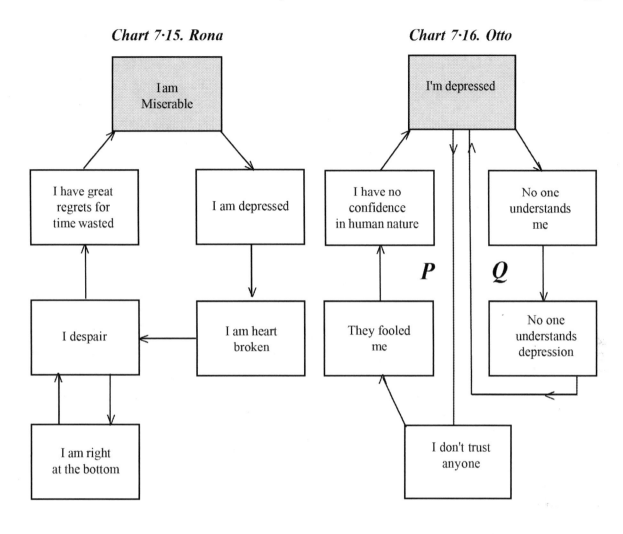

Chart 7·15. Rona

Chart 7·16. Otto

O	seemed to be rather bored and focused on being fooled.
F	Does everyone, all the time, fool you?
O	No, of course not. It's the ones who do that hurt me. Of course the ones who don't hurt are OK. I trust them and human nature is OK with them.
F	What is the balance for you, then, do you trust more people than you distrust?
O	There are far, far more that I trust. It seems silly that I give so much attention to the others.
F	And depression? What is your understanding now?
O	OK, really. You seem to understand me and you understand depression.
F	I don't understand depression. You seem to understand **your** depression very well.

Otto stayed for three more weeks and annoyed several group members by vigorously encouraging them to "Chart out your life." There was probably some relationship between the factors in circuits **P** and **Q** but this was not obvious and it would have been intrusive for **F** to have inferred this in absence of reference to it by **Otto**.

Ruby, Chart 7·17

This young woman presented herself as depressed. She emphasized her relationship with both negativity and lack of hope. Her protestations about the meaninglessness of life seemed oddly inconsistent to **F** because she smiled as she spoke.

Her weak point in the circuit, ready for attack, was "There's no hope."

F	You could own that statement.
R	I have no hope.
F	Hope for what?
R	[After some silence] I do have some hope. There's a guy I like who fancies me. If I attend to what is going on I could get on better at work. My flat's OK.

She went on for some time sorting out positive and negative hopes. She concluded that she didn't need to tell people about her negativity and would enjoy telling about her positive hopes. She experimented with this in group during two further weekly sessions.

Ruby spontaneously illustrated the value of reformulating negative statements into positive forms.

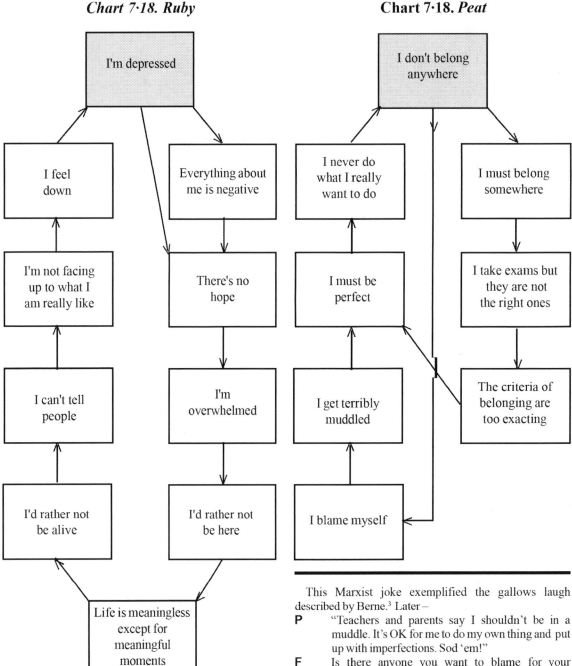

Chart 7·18. Ruby

Chart 7·18. Peat

Peat, Chart 7·18

This young man was interested when others talked about depression but did not use the word about himself. He spent time over three weeks constructing his Chart. The aspects of blame and muddle were added last.

P   I get mixed up about what I really want so I study the wrong subjects and then mess up the exams.

When focusing on self-blame, ringing the word in red he gave the Marx quotation –

P   "I would not belong to a club that would have me as a member."

This Marxist joke exemplified the gallows laugh described by Berne.[3] Later –

P   "Teachers and parents say I shouldn't be in a muddle. It's OK for me to do my own thing and put up with imperfections. Sod 'em!"

F   Is there anyone you want to blame for your muddling?

P   Parents, teachers, so-called friends, they were all a mess. I took no notice of them.

F   Are you taking notice of me?

P   No, not really. I guess you want me to do something and that will only make the muddle worse.

A week later he reported that he had not done much during the week except –

P   I meditated on what I really want to do. I have a brother in Canada. I'll go off there and see what I can find.

That was his last session with **F** who thought **Peat** would only take his problems to Canada with him though perhaps he would find a way of belonging in a new environment.

Pip, Chart 7·19

This middle aged man had seen other group members construct Charts of cyclic behaviour and set out to make his own. As is clear from the copy presented here that every venture out lead back to depression. "Poor Old Joe" (a nick name) was what his family frequently said to him when things went badly. As he said those three words he sounded sarcastic and agreed that his family were not really supportive and mocked him.

He chose to focus on "I'm useless" inverting it immediately.

P   I'm not useless. I care for myself. I swim to keep healthy. I give a lot to others, including my wife. I take a lot of crap without much bother. I'm a highly qualified and experienced engineer. I'm a hero, really. I'm intelligent and can be amusing in the right company. I have every reason to feel secure.

This was not enough. He worked on (worried on) for many weeks dealing with these three factors, varying widely from declarations of happiness to despair.

F thought that this circuit investigation was spurious, done only in emulation of others and was of doubtful benefit.

T   So they call you "Poor Old Joe!" What do you call yourself?
P   "Poor Old Joe" – if I think of myself and give me a name. That's it.
F   Are you happy with that?
P   Crap! No! I'm really Joseph. Good Old Joseph of the Rainbow Coat.
F   Go round the group. Tell each person who you are.
    Pip sat impassively gazing at the floor.
F   Did you hear my suggestion? Tell me if you don't like it!
P   Yes.

Pip got up and slowly began the round. After a while he and the people he interacted with became excited as they evidently knew something about a pop song to do with a rainbow coat.

Sybil, Chart 7·20

This middle aged woman listed her negative features with some alacrity. She dwelt on adopting children as a possible way to brighten her life but concluded that this could be terrible for the children if she didn't snap out of her depression.

A week later she proclaimed her interest in sun and warmth, declaring that she would go and live in Southern Spain. She had some friends there and had already visited them.

F   This seems unrealistic to me. Can you finance a move abroad? And what about your husband?
S   Fuck him. I'll force us to sell up and take half the proceeds for a bungalow in the sun. He can come too, if he likes

Sybil stopped coming to group and during a meeting with **F** in the street said she was going to Israel without her husband.

Chart 7·19. Pip

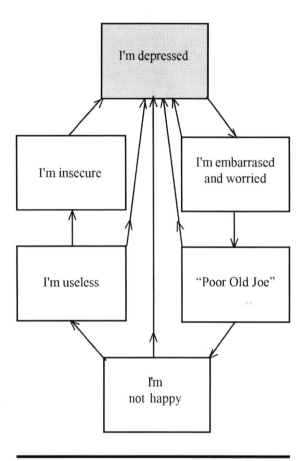

Rab, Chart 7·21

This young man did not raise the question of depression as such. He spoke of being bored and lacking lustre in a very toneless, monotonous and energyless way.

F   As you look at the half-dozen stages of your circuit do you see one that looks more vulnerable than the others?
R   Well, it's my parents, really, they don't actually hold me. I'm so miserable around the place they would probably be glad to get rid of me.
F   Would you like to investigate your relationship with your parents by role playing?

Rab did so and expressed a lot of anger directed, most particularly, towards his mother. During the next three group meetings he reported improvements in "managing his relationships at home" and had decided that it was much to his advantage, particularly financially, to continue to live there.

Rex, Chart 7·22

This middle aged man linked anxiety and lack of sleep saying that he was perpetually tired. He constructed his Chart over two weeks before declaring it to be finished.

F   Where would you like to give your attention first?

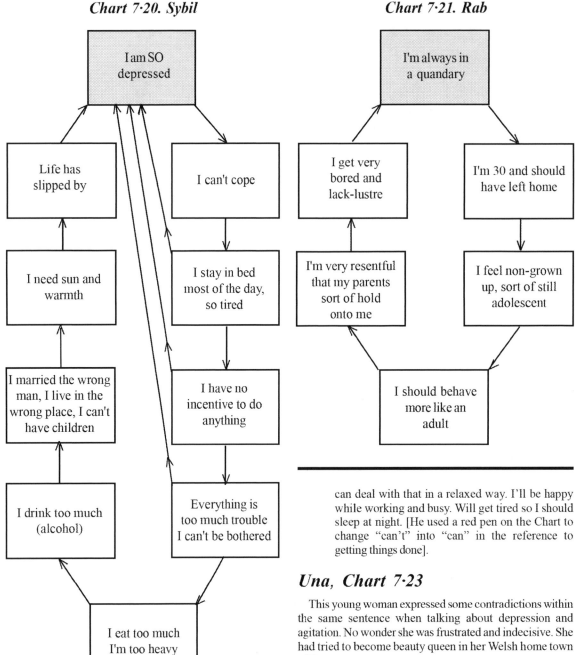

Chart 7·20. Sybil

Chart 7·21. Rab

can deal with that in a relaxed way. I'll be happy
while working and busy. Will get tired so I should
sleep at night. [He used a red pen on the Chart to
change "can't" into "can" in the reference to
getting things done].

Una, Chart 7·23

This young woman expressed some contradictions within
the same sentence when talking about depression and
agitation. No wonder she was frustrated and indecisive. She
had tried to become beauty queen in her Welsh home town
with no success.

F   It seems to me that depression, frustration and
indecision are inactive states. Yet you also say you
are agitated and restless. No wonder you say you are
split.

U   Yes, I did. So?

F   Would you like to deal with the contradictions you
expressed?

U   No, I won't be able to do that. I'm not depressed and
agitated at the same time.

F   You have seen other people deal with contradictory
aspects of themselves as set out on their Charts.
Would you like to do the same?

U   No, I would not. I would be a spectacle in front of
these people.

R   The tension in my body.
F   What do you do about the tension in your body?
R   What can I do?
F   Would you like a suggestion?
R   Don't go on about taking exercise. I have no time.
F   Would you like a suggestion?
R   OK. Let's give it a whirl.
   F then took **Rex** through the de-tensioning procedure of
Grove[1] detailed here in chapter 9. Then –
F   Have a look at your Chart again. What do you
notice first?
R   I notice my emphasis on the big new contract. Yes, I

Chart 7·22. Rex

Chart 7·23. Una

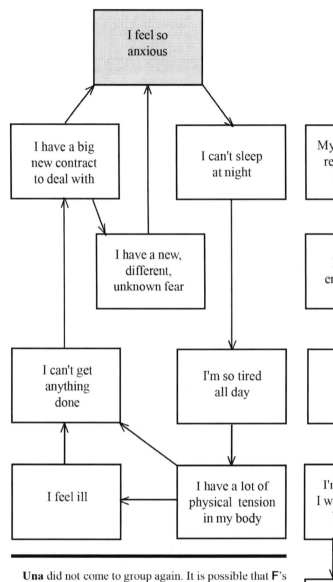

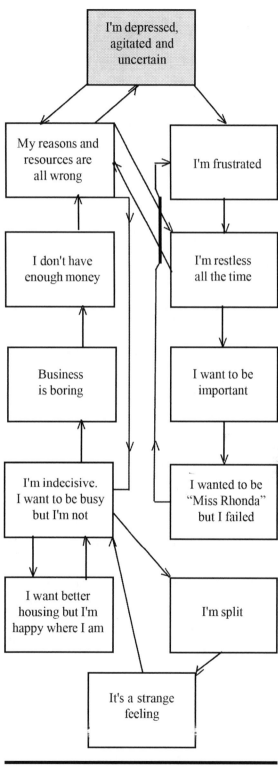

Una did not come to group again. It is possible that F's language was too brief and high faluting so that Una did not feel safe. On the other hand the contradictions she expressed were such as to lead to self-frustration.

Ryan, Chart 7·24

This young man spoke with a grave, learned demeanour, and gave a mini-lecture on depression. The Chart has been constructed from aspects pertaining to Ryan rather than to his book knowledge.

F   As you look at your Chart do you recognize one of your stages as being weaker than the others?

R   Yes, it's this business of doing things for myself. Of course I do a lot for myself though I could do more if I really concentrated on it.

Ryan went on about this for sometime and then suddenly –

R   I'm mostly fed up about the self-pity bit. How ridiculous! How debilitating!

F   Would you like a suggestion? [Assent given by head nod]. Set up two positions, one for your self-pitying

Chart 7·24. *Ryan*   ***Chart 7·25. Unity***

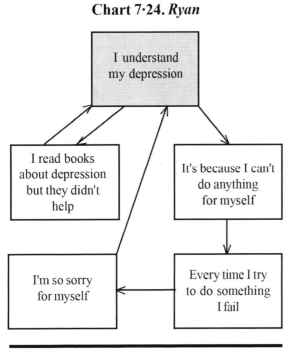

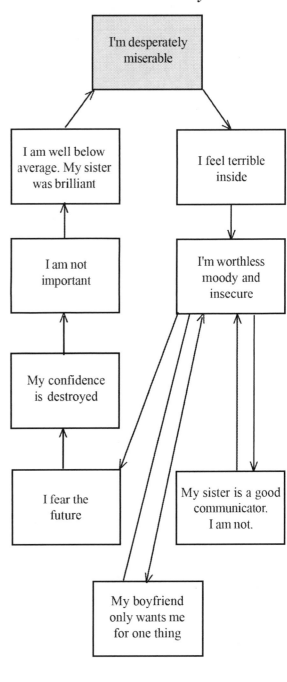

self and the other for your O.K. self.

Ryan engaged in dialogue in the usual way and eventually undertook to monitor the things he did for himself. Over the next few group meetings he enjoyed reporting success in his undertakings and growth of self-confidence.

Unity, Chart 7·25

This woman in her late twenties, presented as being miserable and worthless and was, like **Rex** (Chart 7·8) suffering from lack of confidence. She had many self-deprecating factors active and was strongly aware of inner feelings of badness. She tended to compare herself with a successful sister.

F   Would you like to concentrate on your terrible internal feelings?

U   No. I would not. I would waste your time.

F   I hear that you are not important enough to take the time and energy I offer you.

U   I am important enough. I just don't want to be poked about.

F   I notice that you have just declared your importance while on your Chart it says (points at her Chart): "I am not important." What is going on for you that you express contradictory statements like that?

U   [After some silence]. I don't know.

F   You can tell me what you do know.

U   Sometimes I'm OK. Sometimes not.

F   Say some more.

U   No. [pause]. I guess I tend to fix on the non-importance due to past events that went wrong. I am generally important [looks at the Chart] I am

generally worthy and don't need to feel insecure. The future can be OK.

F   What would you like in your OK future?

Unity spoke for some time about being happy with her boy friend's intense interest and dismissed the relevance of her sister's character. She said she was determined to put her depression in her past. The dismissal of her sister seemed too easy and she worked on their relationship in subsequent weeks.

Chart 7·26. Rory

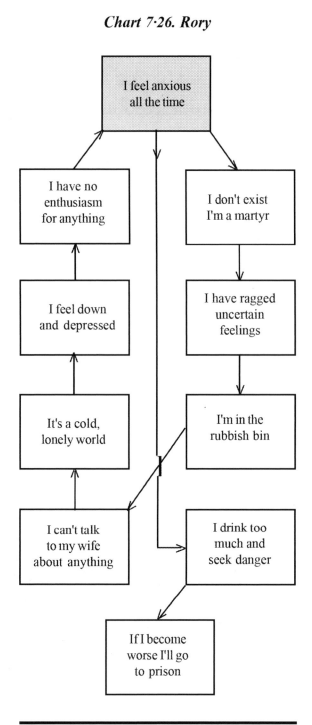

| R | No. She talks to me, I don't talk to her. |

R No. She talks to me, I don't talk to her.
F What do you gain by not talking to her?
R [Energetically] I don't know what you are getting at. I'm concerned about drinking.
F OK. Let's return to your Chart. [points at it] You know about finding a weak point on your circuit; where would you like to give your attention.
R It's wrong about enthusiasm. I have great enthusiasm for a drug, alcohol. It's right about martyrdom. I like to think that people punish me but really I punish myself. So I am not a martyr. That's a fantasy I've always had of being famous by getting things messed up. The world is not really cold – I make it cold. Fat lot that does for me. I make myself lonely too. It's time to warm things up [He looked up expectantly and around the group].
F Who shall warm things up for you?
P (**Poppy**) Oh no mate, you do it for yourself.
R [After much silence] I make myself cold and lonely so I should be able to make myself warm and … (evidently searches for a word) appreciated.
F Ask for what you want.
R **Una**, do you like me?
U Yes, you're OK. I don't exactly love you but you're OK.
M (**Marty**) I like you as you are now, not whingeing, Doubtful, apprehensive perhaps and vulnerable. OK.
U Cold fish you are not.
R [Quietly] Thanks.

Some weeks later –

R I remember feeling anxious. I'm more confident now and drink less. But I'm still not really talking to my wife.
F It takes two to tango.

F evidently expected **Rory** to refer to his mother as the one who would warm things up. **Poppy** was much more to the point. **F** regretted the flippant end remark. **Rory** and his wife went to another therapist.

Comment

These twenty-six experimenters have produced about 100 one-liners which are similar to those quoted by Rowe.[4] Just the same the general quiescence and support of the group situation gave space for the various neocortices to express themselves.

Moving among many people outside the group therapy situation I watched for cycling phenomena similar to those reported here. Occasionally there were signs of such events as with the widow who grieved the loss of her husband. She had a cliché phrase she used to shake herself out of her misery: "This won't get the baby bathed!" And she got on with something that interested her.

Rory, Chart 7·26

This young man evidently temporarily solved his anxiety problem by drinking alcohol and was afraid of over doing it. **Rory** made his Chart and elaborated graphically on his position in the rubbish bin, listing companions as old, smelly tin cans, dirty packets, stinking bones and other obnoxious things.

F If you are in among that stuff there's no wonder your wife won't talk to you.

DEALING WITH DELETERIOUS CYCLES OF BEHAVIOUR

The dialectic neatly expressed:
"The patient is a trainee therapist."[5]

Twentysix stories told above provide the basis for making major and minor generalizations. The first point to emphasize is that each person did not like what was going on for self on joining the group and wanted change. This was sometimes overt, informally negotiated as a change contract. Otherwise I presumed that that was what was wanted as the person dived into a tale of despair and trepidations. Before proceeding I want to emphasize what I wrote on page 113, the stories as told there are grossly distorted by truncation. The investigation of the situation of each person actually spread over many hours of group interaction, during several weeks.

These recycling events occurred in a way that seemed to mimic grounded gestalt cycles but were in themselves not grounded and did not conform to the kiltered pattern as illustrated in the Frontispiece. The general effect of such cycling was to leave the person in suspension.

Whilst the person is out of an organismic, grounded circuit he or she is out of engagement with and dissociated from any other person.

Depression

While this deleterious preoccupation continued the person, bewitched by depression, was effectively at an impasse of the type described by F. Perls who evidently noticed some form of recycling activity when he described people "going round and round" at the impasse[6] (page 92).

As each experimenter bemoaned existence, speaking from an alienated state, staying with an-organismicity, or, as I like to call it, being un-kiltered, seeking a better way of living. As I wrote on page 113, I was struck by the way events followed one another, repeating endlessly, rather like a parody of the stages of the organismic cycle (Chapter 2 and summarized in the Frontispiece). Since recognizing cycling, every experimenter who was imprecise and rambling in presentation was investigated for cycling. However, the contrast in character of the stages was dramatic. Whereas the healthy cycles consisted of stages of generalized affect, engagement, action, satisfaction and eventually becoming grounded, the an-organismic cycles consisted, as detailed above, of a particular moan, with statements about lack of confidence, inability to trust other people, depression, etc., with no sign of grounding.

This recognition lead to a guess (hypothesis) that one stage of an ungrounded cycle would be more susceptible of change than the others as set out in the account of the index experimenter on page 114. A second hypothesis was that if such a change occurred it would be likely to be in the direction of a jump into a healthy cycle. This would then be a curative change. This turned out to be so even if concentration on the matter was suggested by the facilitator.

Once the experimenter was interested in the processes of chains of events, discovery flowed naturally, sometimes spread, as said above, over two or three group sessions and the incomplete Chart of one sessions was produced in subsequent sessions until the experimenter pronounced the Chart complete.

It must be remembered that the experimenter already "knows", in or out of awareness, about organismic cycles; he or she will have traversed many million of them before coming to therapy. As confessed in the last chapter, a simplified representation of the Frontispiece was displayed on the group room wall. Also the action of discovering an organismic cycle was itself organismic and a model for self-nurturing, self-approval, self-improvement and self-esteem.

The detection of deleterious cycling stages; lack of grounding

Some experimenters went through a cycle unprompted, and as stressed in Chapter 3, recycled until stopped. Others completed the cycle with prompting from the facilitator who asked leading questions containing the word "happen."

The Charts in this chapter show slogans summarizing what went on and these were largely the product of the facilitator, though not always since experimenters sometimes produce what are effectively headings for what is going on for them. It seemed to the facilitator that these slogans were adequate summaries. Each experimenter continued exploration in as much detail as the experimenter wanted, taking as much time as necessary, until there seemed to be a pause point when reformulation and other activities could be explored.

Then the experimenter could read the factors around the chart, or the facilitator did, thus adding auditory appreciation to visual appreciation. This multi-focussed process, if conducted by the facilitator was also trance inducing[7].

As I emphasised above the experimenters were, to a large degree, selected. They saw the working out of the cycling phenomenon by fellow group members. In the group therapy situation one experimenter produces his chart of deleterious stages watched by the other group members who are, while attending to what is going on, also aware of their own processes and are influenced to emulate. I neither encouraged nor discouraged such identification They are experimenters for whom I had, as I sat down to write this book, audio-tape recordings as a complete record. As stated above, at least as many as those reported took home the charts they made so that I did not have a full record and these are not reported. My impression is that this lost group probably differed little from the reported group. While I was experiencing the unfolding of the cycling observations I had no intention of writing them up. If I had done so I would have been more careful to keep full records.

In the usual way of a scientist I discovered conclusions and generalizations as I went along, The most important among the latter appeared immediately with the index experimenter. My interpretation was that each experimenter had two trains of events going, one was the deleterious stages of a cycle which constituted "depression" and the other was the desire to return to a more effective state of living. These are dialectically contradictory so I expected each experimenter to spend much time exploring intermediate ground. Although some experimenters effectively did this over several group meetings the majority responded quickly when answering some form of the leading question: "Which of the stages of your cycle is the weakest, – where could you most easily make a change?"

The depressed person has some self-regard but is locked in isolated uni-polarity, self-importance as Castenada says, reacting obliquely to one stupidity and falling into another.

While the experimenters were trapped in cycles that were not organismic, and not grounded, as far as I can see there are no other generalizations available from the data. Each presentation was individually self-crafted to suit the personality of the person concerned. It will be beneficial here to review the situation of these people, one by one, with respect to general concepts of gestalt theory.

Circuits of figure stages as metaphors

The very presentation of the person in group in some way mirrored his or her cycle. It was not exactly his cycle made manifest but a shadowy enactment in which the cycle was effectively a metaphor.

Thus, with hindsight, **Jan**'s gazing at the floor was the enactment of "Nobody notices me" though in the form of "I don't want anybody to notice me"

Leo, avoiding engagement with people in group was avoiding the possibility of an aggressive attack of the style he had from his father.

Lord presented a cycle which was full of failure. His demeanour in group was just like his cycle, his hesitancy fitted with his representation of himself as being immobile because he feared error.

Olga likewise depicted in her cycle "loss of concentration" and she exemplified this in her wandering from subject to subject without completion.

Poppy could have been expected to anticipate what **T** wanted but there was no sign of this.

Marty wanted so much that she didn't get – it is doubtful if she got much from the group.

Max. As far as **T** was aware nothing was stolen from group members or **T**.

Roma particularized her "fed up with people" decision in the group by her quarrelsome relationships.

Neal presented himself as perpetually rehearsing for the real thing. It was the facilitator's persistent questioning that lead to a degree of focusing.

Nuda presented himself with considerable energy. It was not the "I'm paralysed" injunction at work but more likely the "I'm panicky" injunction.

Una frustrated herself by becoming silent as she set out to talk about herself.

With other staged cycles the metaphorical effect was not so clearly visible.

The point of departure from kilter and self-nurturing

The outcome for each experimenter was, as far as I knew, beneficial. I lost contact with most of them, with the few exceptions detailed above, and some of them sent their friends and relatives to group, probably the best indicator of success.

There are times when polarity conflicts are put on hold and may appear to be in a state of equilibrium, dormant and of no consequence. I emphasize this occurrence since otherwise it may appear as if I am saying that life is a constant turmoil. Many married couples find a source of conflict and, having partially worked over it, agree to put it aside; sexuality issues are frequently involved. They make what is effectively a verbal contract to never mention it again and this state of affairs may last until the death of one of them or it may irrupt again as a quarrel after some years. Each of the partners will no doubt remember the contract from time to time, think about it and perhaps find some way of expressing the otherwise suppressed emotion, perhaps to a friend or counsellor. Each has mental energy preoccupied with not breaking the truce, with maintaining the peace but energy which cannot be expressed in the full spontaneity that can be shared joy for the couple.

The paradigm, aggression ↔ assertion ↔ passivity, page 28, is the easiest polarity to spot. **Leroy**'s expression of anger and expectation of anger from other people was very much presentation of confusion due to simultaneous activity of aggressive and passivity factors. Likewise with **Lou** where **Rab** picked up the aggression, expressed as bossiness, although there was much passivity expressed by the experimenter about himself by using the words ignorant, useless, abandoned and depressed.

Each stage in the deleterious cycle of events produced by each experimenter was in contradiction to what that experimenter really wanted for him or her self. The dynamic lay in the working out of the contradictions often lead by the facilitator's question: "What is the opposite of the situation you are concerned with?" In general the facilitators task is to lead the experimenter to face the contradictions: the experimenter makes his or her own clarifications and conclusions. Concentration on this simple emotional issue seemed enough to change Luo's personal environment. Without much attention she stopped insulting herself, losing the words "useless, lazy and stupid." She forgot about shyness and lack of trust and got on with a more positive life style.

With **Otto,** like **Olga**, trust and lack of trust was the main polarity issue with hurt or not being hurt by others as a subsidiary matter that she did not choose to deal with. Similarly **Ruby** chose to deal with hope and lack of hope and Unity chose importance and non-importance. For her other issues could have been worthlessness and worthiness or confidence and lack of confidence. These factors seemed to lessen in importance as she recognized her own importance to herself.

Peat was unfocussed in his muddled state. He reacted to **T** as a new source of muddle. By starting studies (each a figure in itself) that he was not really interested in he set himself up to fail. **T** thought he would only take his problem to Canada with him or perhaps he would find a way of belonging in a new environment.

Arising from stage <1>. **Rex** expressed an-organismicity with his "Life is just a dress rehearsal" slogan. This is not figural and stages beyond <2> are missing: he awaits discovery of interests, appetites and curiosities. One may say that he is un-motivated. He chose to assert himself and found ways up the organismic stages.

Unity's misery and feeling terrible set her at stage <2> where she was unaware of basic needs and interests. She began to move when she acknowledged a degree of importance for herself.

Rory fixed himself among the rubbish indicating poor quality of figure emergence. Add anxiety to this and there was no wonder he was ineffective at doing anything. The facilitator concentrated first on the rubbish, bringing about a swift reaction, negation of some of the factors on his Chart and owning responsibility for his state of being.

Arising from stage <2>. **Jan** began out of engagement, with no energy yet knew what he needed since his complaint was about lack of family and friends. So he began his an-organismic cycle from stage <2>. His panic episodes were pseudo-action stages <5>, with no outcome other than reversion to the beginning of his deleterious cycle; isolation and friendlessness. (The pseudo-action was ineffective because unsupported by energy and real engagement). The reported group session began with the facilitator inviting **Jan** into dialogue and thus into an organismic figure. **Jan**'s subsequent figures were of the same character, even if a bit hesitant in presentation.

Leo, on presentation, was either energyless while knowing what he wanted <2> or talking with moderate energy <3> and engaging only with the facilitator. Therapy consisted in making or moving towards engagement <4> with other people. Having engaged he conducted his business with the person <5> and got on with whatever interested him next.

Olga talked about her experiences with no attention to her listeners and with little energy, presumably at stage <2>, especially when she was expecting people to be interested in her talk about "important things." She found that "small talk" lead her into engagement and action with people again. This change was her choice; she evidently instinctively knew that this would work for her. Her previous choice, church practice, was ineffective in the long run, presumably because it made no lasting impression on organismic figure generation. Each new figure tended to go off an-organismically rather than to completion.

Otto was interested in being fooled and not trusting anyone which he expressed <2> without, at first, much excitement and with no interpersonal engagement. When he got going he focused on the trust issue and got into a healthy figure as he found positivity.

Arising from stage <3>. **Lou** seemed to block off at stage <3> finding little energy until he turned on himself. It was interesting that he saw himself as "busy" and it was

another group member who show bossy words which facilitated easier engagement with others.

Roma was quarrelsome and tended disengage before being active with another person.

Arising from stage <4>. **Lord** evidently knew what he wanted to do but failed in action <5> so his divergence arose at stage <4>. For him "getting things right" was crucial, bringing back into organismicity at stage <5>. This was his choice. He was probably not directly influenced by the organismic cycle which was on a flip chart near by.

Rab was stuck with engagement with his parents with whom he shared no action, only negative affect.

Arising from stage <5>. **Leroy** seemed to engage well with people but did not complete a necessary action <5>, expressing his anger.

Ryan was stuck with the question of doing things for himself.

Marty was apparently unable to get a job and then took the action of going to Eastbourne.

Arising from stage <7>. **Nona**'s cycle consisted in telling people about her depression and this was her action <5> but she got no satisfaction <7> from her relationship with people who heard her and who evidently regarded her as a bore. When she was alone and literally talking to herself she ran her deleterious cycle.

Poppy was frequently falsely active, trying to anticipate what other people wanted. This lead to absence of satisfaction and pleasure in what she did.

Una does not appear above because she produced contrary impression concerning her energy state. The facilitator expected that dealing with the contradiction would get her in touch with her energy but this did not occur. It was also not possible to classify **Max** in the above scheme.

Polarity and dialectics

With the exception of **Una**, who appeared to gain no therapeutic advantage, the experimenters presented in Chapter 9 did not present polarity phenomena in their deleterious cycles. They presented them during therapeutic work. As each experimenter expressed some kind of polarity and worked it out. This work out was a dialectical development. The anorganismic cycle thus expressed polarity conflicts in the dialectical manner. The anorganismic cycle was also a polar opposite to an organismic cycle and finding the latter was the curative move.

Part of the character expression of **Leo** was hesitancy as he both wanted to "speak out" as his father had told him together with the demand for silence. Polarly opposed injunctions leading to confusion. Likewise **Leroy** was projecting his fantasies and **Una** was using depression and agitation to support a confused state of mind. **Lou** benefited from experimenting with the polarity –

I am afraid ↔ I am courageous.

Therapy; from deleterious to beneficial figures

The following are types of reformulation changes that were apparent in the above as the affected person changed from dissociation to association relative to other people.

Polarity inversions

These may be simple inversions. Thus **Jan** changed his interest and evaluation from not being noticed to being noticed too much. Other observations are set out in Table 7·1.

Re-evaluation

Here **Jan** who had been concerned about panic attacks had had only one minor episode during one week and had fallen asleep after it.

Leo said clearly that he wanted to be able to "speak happily to anyone and everyone." He slowly achieved this in group and outside though it took a long time.

Olga found her escape route by herself at home when she experimented with small talk.

Rab realised that his parents were not holding him, he was holding them.

Ryan gave up self-pity for self-esteem.

Otto's hidden injunction was: "No one can (or will) understand me". "Don't trust anyone" and "I won't let them fool me."

Peat "I don't belong anywhere". "I do the wrong things," "I must be perfect," "I blame myself," "I muddle myself."

Sybil "I can't cope," "I have no incentives," "I can't be bothered."

Unity "I'm always miserable, feel terrible inside, worthless, bad communicator and have no confidence."

Rory had a complex story about being a martyr that he has probably picked up as a child. he certainly felt rubbishy as well as lonely and depressed.

Change of focus

During the main body of the work with the facilitator **Jan** had been reminiscing, probably talking to himself rather than to the facilitator. He changed his focus at the facilitators instigation, to talking to first one and then other group members making the beginnings of warm, organismic engagement.

Olga's experiment with small talk lead to a change in which she began to relate warmly to her family. It was as if her decision to avoid small talk alienated the affections of her family.

Unity changed focus from terrible internal feelings to her "boy friends intense interest."

Specificity issues

Reformulating from vague statements, often involving "it" or another relatively unspecific word, to the particular leads to greater understanding of what is going on for both experimenter and facilitator. Thus **Ruby** wanted to "show her." It was already known that "her" was her mother and she became clearer about what was going on when she recognized that she wanted to show mother "I am boss" and "I will decide what to do."

Nudd's "things are always wrong for me" when challenged for specificity transformed the meaning of "things" and "always."

Otto being fooled and misunderstood achieved some sense of balance on balancing foolers against the others.

Lacunae

Jan, in proclaiming that nobody noticed him, was expressing himself about a blind spot and he opened it up for himself when he talked about being noticed too much.

Nona's lacuna was a non-functioning memory which responded to techniques for improving memory.

Poppy didn't realise that the anticipation game she needed to play when young was not necessary and was debilitating at her present age.

Neal's lacuna was to do with the real theatre of life and originally he didn't have any concepts as to what this might be.

Otto lacked hope. **Rory** effectively did not exist.

Reformulation of negative injunctions

Injunctions are decisions made in the past which become instructions to the self and continue to act as such. Thus teachers induce a teenager to affirm: "I'm stupid" and then as an adult this injunction poisons every attempt to do anything. The essential, therapeutic move, as the Gouldings[8] say, is change the negative attitude to a positive one, negate the negation as I say.

These injunctions all had negative, deleterious effect and the descriptions of the negation of these effects is obvious and not described. The great advantage of this technique is that a gamut of factors is exposed and the experimenter chooses which one to deal with rather than jumping on the first one to come up. Negate negativity to obtain a new positivity; how obvious.

Sometimes the injunction is expressed directly and sometimes hidden, usually in dissociated language

Jan presented no clear injunction. "It's so depressing" is in dissociated language and hides the real injunction: "I'm depressed."

Leo could have more clearly said: "I embarrass myself and make myself shy when I try to speak to anyone. This stops people treating me like my father treated me."

Leroy characterized himself as basically arguing with all people and things.

Lord accepted his mother's view that he was never successful with jobs. She probably began to present that injunction to him when he was very young.

Nona's pungent message was "I can't remember anything." Maybe mummy said something like: "Poor dear, you never remember anything."

Olga's "I lose concentration" was a mild form of "I can't (or won't) concentrate So I keep on worrying and become despairing."

Pearl told herself she was "useless, lazy, energyless, without stamina, shy, readily embarrassed, stupid and couldn't trust herself." What a barrage of negativity! She chose trust as the issue to contend.

Poppy rejected criticism.

Neal 's statement "Life is just a dress rehearsal" was his excuse for procrastination, waiting around, McCawber like, for something to turn up.

Nudd's destructive injunctions were, "I'm paralysed, I panic easily, I get upset, I want to vomit, and I'm useless."

Rona's were similar; "I am depressed, I am heart broken, I despair, I am at the bottom, and I have great regrets."

Otto's hidden injunction was: "No one can (or will) understand me". "Don't trust anyone" and "I won't let them fool me."

Peat "I don't belong anywhere". "I do the wrong things," "I must be perfect," "I blame myself," "I muddle myself."

Sybil "I can't cope," "I have no incentives," "I can't be bothered."

Unity "I'm always miserable, feel terrible inside, worthless, bad communicator and have no confidence."

Rory had a complex story about being a martyr that he has probably picked up as a child. he certainly felt rubbishy as well as lonely and depressed.

Double binds

It is of interest to consider the role of double binds in the aetiology of these experimenters, although this account is all hindsight.

Jan focused attention on "nobody notices me" and then spoke of people noticing him too much. With **T**'s help he attended to his relationship with **Netta** and found appreciation for her concern. When he said "People I like notice me" he was reformulating his blank "nobody notices me," which, so **T** thought, implied "I don't want to notice anybody."

Such double binds as he had dissolved on abandoning generalizations and particularizing with **Netta**; he soon stopped seeing her as an *it* and found her as *thou*.

Leo had a clear double bind from his father be silent and speak out in the world.

Leroy was angry with people, expecting them to be angry with himself and was surprised when they were not. On role playing angry people he produced a positive bind which he found to be funny.

Lord both blamed and obeyed his mother until he became clear about what he wanted to do for himself.

Olga was unable to concentrate on most relationships and yet connected to important things, as she said, like Wittgenstein. The bind seemed to be around appreciating only the very serious side of people. As she loosened up and chatted more about "trivia" she engaged better with people.

Pearl had evidently had "you are stupid" messages when a child to which she reacted by "trying hard." One pole from

parents and the other from herself, with the general effect that she failed to get appreciation from anybody. She found her own way by concentrating on being courageous.

Lou did not realize the effect his bossiness had on other people. He felt the attention of other people and then lost it as they revolted against him, leaving him confused and angry.

Poppy seemed to be in the grip of an "I'm wrong if I fail" and "I'm wrong if I succeed" dilemma.

Marty seemed to be unable to deal with the messages "I can't have what I want" and "I'm wrong if I get what I want."

Max had messages to do with "It's wrong to steal" and "It's exciting and fun to steal."

Roma. T supposed that part of the problem here was that **Roma** gave other people advice which they did not heed. This rejection fed into her feeling like a child, so the bind here was about not being appreciated as giver or receiver of advice.

Neal defiantly opposed his Mother's interests. There might have been a bind in this if, at some time in his past, she had criticized his way of being tidy but **T** did not pursue this.

Nudd had ideas that he used to immobilize himself and set off panic attacks. There seems to be messages about being wrong for "being useless" and wrong for endeavouring to do things.

Otto had a dialectic between "you don't understand me" and "I don't understand you" which was resolved by attending to specificity.

Ruby had the concept of "there's no hope" which seemed to contain two messages "I'm not OK if I rely on hope" and "I'm not OK if I have no hope."

Peat had the "I don't belong" and "I must belong" dilemma as his weak point but **T** decided not to make his muddle worse.

Pip. T thought that there was some kind of bind here because **Pip** readily alternated between positive and negative expressions without resolution.

Rab had a polarity concerned with "being held on to by his parents" and "the need to hold on to his parents." The bid from them was "It's OK if you stay" and "It's OK if you go" with no feeling on either expression.

Rex probably suffered from "It's wrong to sleep at night because I must deal with my problems" and "It's wrong to stay awake and be tired all day next day."

Una seemed to be labouring under "It's wrong to be important" and "It's wrong to be un-noticed."

Ryan had "It's wrong to pity my state of depression" and "It's wrong to do anything about it, like read books about depression."

Unity had "It's wrong to be important and confident" versus "It's wrong to be unimportant – especially when my sister is so important."

Rory had "It's wrong to martyr myself" and "It's wrong to be enthusiastic."

In the event all these double binds were sorted out without direct attention to them so they are not of great importance in the therapy situation.

Table 7·1. Polarity changes for the experimenters

Client	Presented observation	Derived observation
Jan	Accused G of intrusiveness	Appreciated G's care and attention
Leo	I can't speak to anyone	I want to speak happily to everyone
Leroy	Anger with himself	The ridiculous ineffectiveness of anger
Lord	Making mistakes	Getting things right
Nona	I can't remember anything	I remember that I can't remember
Olga	Disliked "small talk"	Enjoyed chatting to friends and relatives
Pearl	I don't trust myself	I can trust myself and be courageous
Lou	I am bossy	I can avoid bossy words
Neal	Life is just a dress rehearsal	Life is the real play
Nudd	I'm useless	I am really capable
Rona	I'm depressed	I live
Otto	I don't trust anyone	I trust some people
Ruby	There's no hope	I do have some hope
Peat	I blame myself	I can meditate on what I really want to do
Pip	I'm useless	I care for myself
Sybil	I need sun and warmth	She went to Israel
Rex	I can't get things done	I can get things done
Unity	I'm not important	I am generally important and worthy
Rory	I make myself cold and lonely	I can make myself warm and appreciated

Ancillary aspects

While each non-cycling experimenter tended to block at a specific point on his organismic cycle, the ensuing therapy process involved different, new figures. The contribution of the fresh cycle was largely as a memory process – the activities described were not, with few exceptions, happening in group. Each was, however, an organismic therapy cycle. Moaning about despair <2> was followed by enough energy to engage <4> and get on with the action of talking about the deleterious cycle <5>. Ultimately satisfaction was expressed at completing this unit of knowledge.

Having completed a circuit a certain amount of rumination occurred, new circuits units with little energy, verifying data. Nine aspects of deleterious behaviour have been discussed above. Here are several other minor aspects that are familiar in gestalt therapy that I will now give attention to.

Energy. **Rory** showed dramatic energy change on being challenged about not talking to his wife. This made a great difference to possible outcomes when dealing with the factors on his Chart. The energy change indicated involvement with stage <4> and lead to engagement with people.

Exaggeration. The presentation of information in group is all selective and exaggerated. By emphasizing the exaggeration in the way that F. Perls[9] advocated, the polarity appears followed by resolution. In the example of **Otto** it soon became apparent that he was not being fooled by everyone, all the time. Likewise **Ruby** reformulated "there's no hope" into a positive statement about a man friend.

Back tracking. Ask an anxious person: "What happened before you felt anxious?" and the answer will be accompanied by increased excitement, even if only vaguely. Anxiety is partly fear of being anxious, so, go back to being excited. What is the worst thing that can happen to you if you get excited. Mother's injunction comes up: "Calm down, you will only end in tears."

The gain challenge. **T**'s "what do you gain by ..." question was effective with **Rory**. It could have been used with **Neal** for whom it seemed as if the "dress rehearsal" statement followed some thought or memory that **Neal** had had. This, in the event, was without outcome as **Neal** got on with something else that concerned him.

Dreams. Most dreams are clearly nothing to do with psychopathology and I agree with Dave Dobson[10] that most dreams are in fact excretion of unwanted mental impressions and have no more significance than that, even though there may be highly charged emotional content. Exceptions are dreams which do not reach completion and may then recur.

As the present work proceeded it became clear that, indeed, the main characteristic of a remembered dream was, as Perls said[9], lack of completion. Moreover, the abrupt ending of figure development occurred at a specific stage of the cycle (Frontispiece). In my experience the most frequent stop point was at engagement <4> – the dreamer knows what happens next but does not do anything about it <5>. The common example is the falling dream in which the dreamer is about to crash onto something. Catastrophe is expected. This is an engagement without action hypothesis, as illustrated by **Lord** (page 116), where, in his dream, he wanted to jump across a chasm between fire escapes but hesitated in fear. Completion of the dream occurred while waking by a method that was simple, even naive, to the point of absurdity – but this didn't happen in the dream.

Therapy,[11] if necessary as with a recurring dream, consists of completing the dream as Perls prescribed. The plunge in a dream is invariably, in my experience, not followed by a catastrophe but by something completely different, a gentle descent onto a floor or ground or a simple splash into water, any of them a happy ending. And that is the end of the matter though each dreamer had to discover this for him- or herself.

Another style of dream came up with group members, first one and then the original capped by others with their impressions. Further investigation with students showed this style of dream to be relatively common. **Seb** reported: "I needed to find a toilet to shit but every one I found was occupied with doors open and someone sitting on the seat. I soon awoke with strong urgency and left my bed to go to the bathroom." Variants of this dream involved urination and hunger. **Vesta** said: "I was very hungry and surrounded by people who were eating. None of them gave me a bite and when I woke up I was indeed very hungry and went to the kitchen for breakfast." Attempts to analyse these dreams by identification with the supposed projected parts failed to provide **Seb** or **Vesta** with insights that were of significance. No one reported a variant of the dream style involving thirst or other physiological phenomena such as sexual need.

These dreams involve awareness of need stages <1> and <2>, fantasized energy <3> and engagement <4> but, fortunately within the dream, no action <5> or further stages. Waking up or exploration of reality enabled the action and subsequent stages

Many factors are involved in the processes of figures and some of these are summarized in Chart 7·27.

Polarities among group members

When someone becomes angry in a goup other members are likely to become frightened and passive. The knowledgeable ones will strengthen their assertiveness. At feed back time these feeling may be expressed with benefit to the angry one who comes to realise the effect of his power play, like it or not. The relationship of Leroy and Pearl (page 116) illustrates this generalization.

The role of protofigures

Precursors of the presenting features for these 26 people must occur in ground. Thus **Jan** (page 126) spoke first of being isolated and lonely so he presumably has a protofigure that has developed while being alone and depressed. While it seems likely that the protofigure has the characteristics of the presenting statement this is not necessarily so, although it may otherwise have all the characteristics that appeared in the deleterious cycle. **Nona** may have been basically concerned with bad memory rather than depression which is the affect of her protofigure. **Pearl's** protofigure may be to do with shyness and embarrassment with fear as the dominant affect.

For each person dealing with the weakest manifestation in the circuit of ideas and events lead the person to reformulate that one and depotentiate the others. The protofigure had been modified, lost its affect and probably abolished in a process of relationship with the facilitator which was largely intellectual.

And so on again –

Following this summary[12] of matter concerning 26 experimenters comes an essay in which I wonder what are driving me now, the genes I have inherited from a long line of ancestors, the people I live with or both.

> I, as a facilitator
> have abandoned rules and techniques.
> I have a knack.

Chart 7·27. The engagement and action cycle in lack of health: a summary.

Consummation would be with respect to the initial need or want, etc.
At each stage of this algorithm a judgement of success or failure is made; Y = success and N = fail.

This Chart is constructed without over all reference to a time scale.

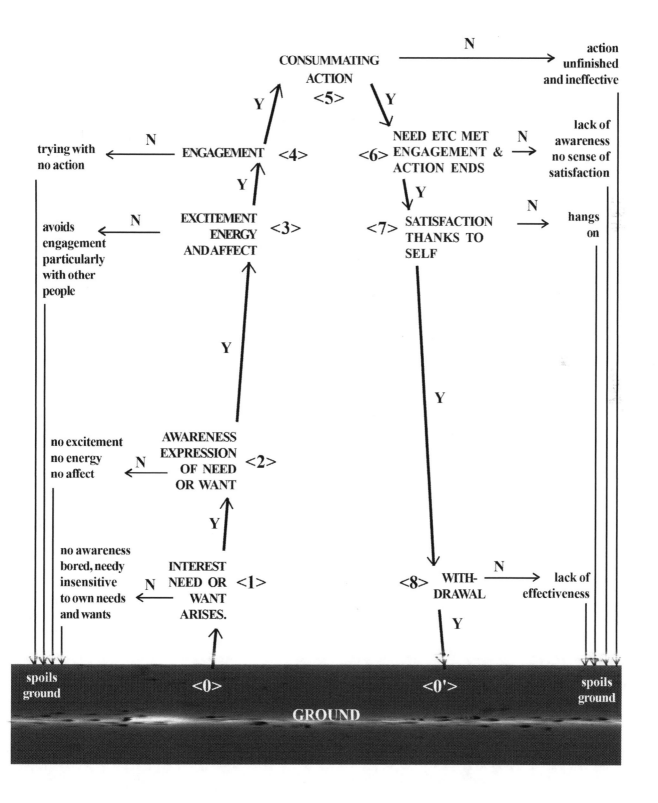

Notes and references

1   Grove, personal communication.
2   Erickson et al. 1976.
3   Berne, 1973.
4   Rowe, 1983.
5   Perls, Hefferline & Goodman, 1951, p. 292.
6   Perls, F. 1973, p. 148.
7   Erickson, Rossi & Rossi, 1976.
8   Goulding and Goulding, 1979.
9   Perls, F. 1955.
10   Dobson, D. 1981. Personal communication.
11   Work on a dream is best carried out in the familiar sleeping posture of the person which is usually flat on back or front or curled up on one side.
12   A quiz programme entitled "The weakest link" has appeared on British TV. This chapter was drafted some ten years before that event.

One likes
to be here, right now
3
for benefit or detriment.

—=(☆)=—

Chapter 8

Gestalt in an Evolutionary Context

The Evolution of Species
is not a theory
it is a fact.
Buss[1a]
And Gestalt Analysis is a Fact.

Any discussion of evolutionary and gestalt psychologies depends on three hypotheses –

- That evolutionary psychology is the over arching, all encompassing concept of psychology.
- That evolutionary psychology needs a theory of perception.
- That gestalt is the basic theory of perception.

While the above points relate to general psychology they can be taken to relate also to psychotherapy. They will now be discussed in turn.

Meanwhile, is the experimenter aware of the benefits available by giving attention to both genetic and/or social factors?

Evolution

Evolutionary concepts
are a source of novel insights
into why individuals behave as they do.
McGuire and Troisi[2a]

Darwin[3] and Wallace surveyed the history of natural species, noticed countless relationships and decided that a long evolutionary chain had occurred. The mechanism was inheritance of acquired characteristics which has been explained and confirmed in our time as passage of chromosomes and the genes on them from parent to progeny[4].

Darwin had been very interested in inheritance in domesticated animals when man controls the processes of un-natural selection. He saw in nature analogous processes of spontaneous natural selection which occurred very slowly during geological time, a span of hundreds of millions of years.

Evolution is a two-stage phenomena: the occurrence of variation and the sorting out of the variants by natural selection. The natural selection process consisted of some creatures being able to deal with changes of the environment,

thriving and changing their traits whilst others did not. A neat example were our remote ancestors who survived in a cold climate by having the intelligence to use animal skins as clothing and being able to light fires.

Factors known to control evolution were –

- availablity of food – our ancestors were predators on other animals and foraged for fruit and roots.
- skill in avoiding being caught by predators.
- success in mating and producing children.
- skill in competition with rivals in finding food and mates.
- skill in cooperation with relatives and friends in tackling large game.
- the climate which, for example, was subject to desert conditions or ice ages from time to time.

There are many other factors, including variation in the needs of women and men.

There is some variation of opinion among experts on exactly how certain phenomena occurred but the consensus is pro-Darwin in general, except for uneducated, unintelligent people who believe that creation occurred about 6000 years ago. It is not only the fossil record in the rocks that substantiate evolution. Many studies among peoples who live in places remote from civilization indicate the truth of the evolutionary story. (Plotkin[5], Dawkins[6]).

Evolutionary observations and ideas are regarded as fact because they meet the scientific paradigms. They also "… organize known facts" and "… lead to new" propositions that are testable and found to be true (Buss[1a]).

The ancient person was pitted against other creatures and other people who were of course part of the environment. While our basic genetic make up has changed little, if at all, during the last 100,000 years, bacteria and viruses in our environment have changed thousand fold leaving us vulnerable to infestations that take all our ingenuity for coping. Tuberculosis was such an infestation and was largely controlled until recent mutations freed it from drug control. (Brüne[7]).

Evolutionary psychology

> Natural selection does not select for health,
> but only for reproductive success.
> Nesse and Williams[8a]

After nearly 100 years psychologists have begun to look at evolution for clues to the origin of some of the dilemmas of psychotherapy. Coolidge and Wynn[9] write –

> The fossil and archeological evidence suggests that there were two periods marked by specially significant cognitive developments, the first about 1·5 million years ago, and the second only about 100,000 millions years ago.

Anderson[10] writes that the problem is –

> … to evaluate the complex interactions of two simultaneous, interactive, but different processes, biological evolution and cultural evolution.

The central hypothesis held by the evolutionary psychologists is that evolution has produced individuals shaped anatomically and physiologically for the life of a million years ago. All mixtures of evolutionary and immediate, proximal influences are not conflicting; symbiosis leads to strong character and behaviour formation. The effects of miniscule mutations are both positive and negative so that, over millenia, an homeostatic equilibrium is established and the character of genetic inheritance has, to repeat, changed little over the millions of years (Schwartz[11]). The undaunted pace of the stone age hunter provides benefit to the modern olympian runner. The contrast of the effect of paleolithic factors and logical thinking on behaviours has been discussed elsewhere[12].

McGuire and Troisi[2b] write:

> "therapy should aim to facilitate [attainment] of the goals of patients … [and] improve how patients regulate themselves physiologically and psychologically."

Important social factors include competition whereby some offspring variants were more successful at gaining advantageous food and mates than others, produced more offspring, and out competed their rivals. Living communally in villages facilitated these activities. There were problems with cooperation in action with groups becoming competitive as the men fought one another for food sources and access to woman. Here there are evidently more polarities.

Evolutionary psychologists rely on evidence provided by archeologists, anthropologists, paleontologists and geneticists and also have highly active imaginations. Anyone interested in the subject is encouraged to have the same although Rose and Rose[13] and the thirteen authors they edit don't agree.

An ancient peasant

George in 1:1 complained of "writers block." He described how "he couldn't think properly" and was "obsessed by a call to his garden," an activity he liked but all the same he was supposed to be writing to a dead line. I suggested that he put his writer self on a chair and talk to him as a gardener. After some exchanges gardener said: "You are not feeding yourself properly." With the reply "You are not letting me think." I intervened: "I have a suggestion for an experiment, would you like to hear it?" After an agreeable nod: "Change your roles a bit. Change the gardener to be a stoneage hunter interested in killing a deer but also eating lots of fruit. And change your modern self and become as sophisticated as you can." He jumped out of his chair saying: "No, I can be a bit crude sometimes but I want to be a writer." We discussed how terrible it must have been to live in ancient times.

George phoned next day to report that he was enjoying writing again. The next week George was concerned about how he could know what he was doing, behaving like an "ancient peasant" or being a writer. I had no suggestions at that time but he thought he could "jump into modernity" by imagining he was Thomas Hardy who "wrote a lot about peasants."

Evolutionary therapy in general medicine

> …natural selection acts
> not for the benefit of the species
> but normally for the benefit of individuals.
> Nesse and Williams[5b]

Human clinical ingenuity can combat the mismatches, as Eaton & Eaton[14] write –

> "For many chronic diseases (osteoporosis, type II diabetes, hypertension, coronary heart disease, lung cancer, and others), the concept of discordance between genes and lifestyle suggests relatively straightforward prevention strategies. In each case, readopting the essential elements of preagricultural life, in terms of nutrition, physical

---

Chart 8·1.   A generalized algorithm for the functioning of the two paths of evolutionary psychology and psychotherapy

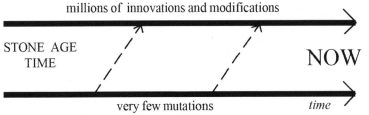

millions of innovations and modifications   environmentally determined behaviour = proximate, contemporary and **cultural**

STONE AGE TIME   NOW   genetically determined behaviour = ultimate, stone age based **biological**

very few mutations   *time*

Table 8·1. Evolutionary mismatches

Stone age behaviour	Consequence of not complying with stone age behaviour	Remedy
The diet of meat & veg was low in fat	Adiposity, heart disease & diabetes type II	Eat a stone age diet
Dietary intolerance to modern cereals & legumes	Allergic type of reaction	Eat a stone age diet.
Neonates had constant body contact	Colic, cot deaths,	Give babies frequent body contact.
Fever with raised temperature	Fever lasts longer if treated with antipyretics	Don't try to lower temperature
Fear of spiders	Reaction to the spider in the bath	Desensitize
Fear of snakes	Unnecessary aversive reaction	Desensitize
Quadrupedal motion preceded bipedal	Back ache.	Chiropractic, osteopathy or shaitsu
Much walking while foraging	Stiff limbs and back	Walk more
Hygiene was primitive and provided maximum immunity	Skin lesions due to loss of natural oils.	Don't be too clean
Wood smoke in the hut	Reduced lung infections	Go for clean air rather than industrial pollution
Chewing poppy seeds produces a nice effect	Ineffective living	Clean living

exercise, and other behaviours, should markedly reduce disease incidence." [the "should" seems to express hope.]

The lifestyle is associated with problems and we use our ingenuity to solve problems; that is what we are good at! The irony is that we and our forbears have created most of the environment that now teases us; global warming!

The examples below illustrate mismatch between environmental and genetic behaviours, polarity situations, creating problems that are resolved by having the affected person choose one or the other behaviour because either may be beneficial.

Evolutionary psychotherapy

As physiology provides the scientific base for medicine
so the evolutionary scheme of things
provides the base for psychology,
McGuire and Troisi[2c]

Travathan, Smith and McKenna[15] summarise the situation neatly –

"Evolutionary medicine takes the view that many contemporary social, psychological, and physical ills are related to incompatibility between the lifestyles and environments in which humans currently live and the conditions under which human biology evolved."

It has become clear that, certainly for people, there are two components in the account of evolution. One is us in our current environment of high technology to the point of space travel – though most of us are content with TVs, DVDs, motor cars and toasters.

The other component is in our genes. They are said to have had only minor changes in the last million years. The outcome is occurrence of mismatches between our cultural influences and our biological influences (Chart 8·1). At the worst these can detrimentally influence our general behaviour which then upsets our social relationships and thus our social psychology. They can also directly influence our psychological attitudes. See Table 8·1.

Nesse and Williams[5c] are interested in psychological problems:

Most of the problems people bring to psychiatrists involve aversive emotions – sadness, anxiety, jealousy, anger, or boredom. [Such] fundamental emotional capacities have been shaped by natural selection because they gave selective advantages.

These authors define these defences as events that appear to be abnormal but are in fact "sophisticated adaptations." "They are observed only when aroused by cues that indicate a situation where they may be helpful." He discusses examples of events and connections and these are amplified, set out in Table 8·2 and show how so "much suffering can be adaptive[5c]." These many events can be disliked but are useful and healthy in certain circumstances. Finding these aspects of suffering to be beneficial is said to be unique to evolutionary psychology. Buss[1b] writes that knowing fear is valuable when faced by a tiger and anxiety can be a valuable warning of occurrence of an adventitious situation. An example of working with two evolutionary aspects is in the box labelled "An ancient peasant".

Nesse and Williams[8] link anxiety with Cannon's fight, flight or (a recent addition) freeze in reaction to danger. Agoraphobia, the fear of leaving home and similarly with other

Table 8·2. Benefits and suffering

"Based on the observations of Nesse and Williams[8c] with my interpretations

Apparent suffering	*Benefit*	*Remedy*
Pain	Indicate tissue damage	Take an aspirin, etc
Coughing	Clears respiratory tract	Take some linctus
Vomiting	Removes toxins from stomach	Drink something soothing
Diarrhoea	Clears toxins from colon	Take some kaolin & morph'
Fatigue	Protects against tissue damage	Flop out!
Fever	Defends against infection	Sleep it off
Nausea	This food is disgusting	Throw this food away
Anger	You are or this is anathema	Cool it!
Sadness	I have lost something I need	Find something else
Grief	Extreme sadness	Mourn
Depression	I have lost everything that was of interest for me	Find something else
Jealousy	You have what I want	See a therapist
Boredom	Nothing interest me	Wallow in it!

phobias are protective against injury and life preserving reactions of ancient origin because danger lurks outside. Therapy helps phobia sufferers.

Social withdrawal, i.e., from environmental pressures, in depression, personality disorders, phobias and others, is an attempt to adapt and alleviate symptoms. (McGuire and Troisi[2d]). Meditation in the quiet gives the mind a chance to evaluate a situation and perhaps provide solutions to problems.

Does the evolutionary attitude really work when applied in therapy? The many anecdotes abbreviated in Table 8·2 indicate success as do the following accounts.

Buss[1c] presented twelve examples validating his observations about pathology and evolutionary psychology. He[1a] also discussed the plight of young men of today who were violent towards their wives and who became more civilized when they realized that they were enacting stone age life styles. Pinker[16] discussed the role of language.

McGuire and Troisi[2d] present 22 reports of patients who worked with them when they used evolutionary psychological principles as their way of working. Only two had failed or doubtful outcomes. Among the behaviours receiving their attention were: the need to change work patterns (changing way of life to improve nutrition), improving kin relationships (the blood of kin is thicker than that of non-kin), to change the environment (when it clashes with dominant genetic determined needs), improved social status (and ability to improve nutrition) and facilitation of the processes for finding a spouse (and thus for producing children).

The evolutionary psychologist demonstrates a genuine need for the paleolithic diet so a gestalt orientated person would probably tend at least to experiment with it. Cordain[17] presents a book of "Paleo Diets"[18] and writes;

You may notice ... that health problems you had lived with or ignored for years begin to improve. Your joints are no longer stiff in the morning and your sinuses are ... beginning to clear. ... For those with ... serious health problems, such as high blood pressure and elevated cholesterol or type 2 diabetes, symptoms begin to improve within weeks of adopting the 'paleo diet'.

We are interested in psychotherapy and as Nesse and Williams[8a] write:

The discovery of a genetic determination for a disorder may provide the best hope for an environmental treatment of it."

Protothings

Gestalt protofigures are neurological mini-modules and as subject to genetic and social determination as any tissue. The sense-phenomena, lumps, butterflies, hedges, aches, itches, clouds hanging over, motor tics, tears, etc., have been around for millennia.

Notes and references

1 Buss, 2008, 1a p 38; 1b p 414; 1c pp 60 → 70.
2 McGuire & Troisi, 1994, 2a p 278; 2b p 281, 2c p 50, 2d 265 → 271.
3 Darwin, 1859.
4 Plotkin[19] finds roots of Darwin's ideas in Aristotle.
5 Plotkin, 1997.
6 Dawkins, 2009.
7 Brüne, 1994.
8 Nesse & Williams, 1994, 5a p 97; 5b p 113; 5c p 353.
9 Coolidge & Wynn, 2009.
10 Anderson in Dunbar, Barratt & Lycett[20]
11 Schwartz, 1999, p 323.
12 Edwards, 1985, pp 84, 149.
13 Rose & Rose, 2000.
14 Eaton & Eaton, 1999. p 430.
15 Travathan, Smith, & McKenna, 1999, p 3.
16 Pinker, 1994)
17 Cordain, 2008. p 212.
18 I have myself obtained considerable benefit by maintaining myself on something like the "stone age diet."
19 Plotkin, 2004, p 14.
20 Dunbar, Barratt & Lycett, 1994, p 334.

—=(☆)=—

Chapter 9

Protofigures as Natural Phenomena

Before a gestalt figure appears
lying deep in a vortex of nerves
is a protofigure.

Everybody feels, senses or knows phenomena like "butterflies in the stomach", "lumps in the stomach or throat", "aches in the head", "black clouds hanging over me", or "a hedge or wall" that seems to appear when approaching a dominating person. For all the appearance of reality these are ephemera and fantasies. The person with such a felt-sensation knows so but nevertheless will tell colleagues about the impressions. These are real unreal events or unreal real events and are natural phenomena for the person knowing them.

It is sometimes obvious that the felt sense is connected to something. A fearful person feels the flutter of butterflies. Another person who suffered a traumatic event knows that the nightmares and/or day time flash backs are connected to that event although, alternatively, the person may not remember the initial event.

Sometimes the felt sense is not connected to anything in particular. A motor tic in the right eye could be an example as the experimenter does not neccessarily make a definite connection to an event in the past.

These events were described by Gendlin[1] who introduced the felt-sense terminology as was stated in the introduction (page 4). Gendlin developed a complicated way of using felt-senses in therapy which was systematized by Grove[2] who discussed events in terms of therapeutic metaphors (Kopp[3]). The therapeutic situation became clearer when the present author identified felt-sense events with gestalt phenomena, identifying each felt-sense as a preliminary event, as a felt-protogestalt figure, for the development of a full gestalt figure.

At this stage it would be possible to describe the felt-protofigure and gestalt events in abstract terms but I judge it to be clearer to describe examples in this chapter and discuss the minutiae of technique and ideas in chapter 10. During work with protofigures the facilitator expresses himself using Grove's "clean language" which is fully described in the Appendix, page 163.

The first two examples are concerned with the full proto-figure therapy procedure where two felt-protofigures are discovered and manipulated. Then follow descriptions of events where one felt-protofigure was discovered. These are more like the events described by Grove[2].

Dealing with a trauma or problem

Grove's hypothesis (in my words) –
each protofigure is
connected to a trauma or problem.

Five sessions dealing with felt-protofigures will now be described. The stories are synthetic for clarity and brevity but are based on real observations with experimenters who had problems including those they described as traumas.

Elucidatory comments follow each story and their sources are indicated {xN} in the text*. A more generalized discussion follows.

Double drive

In a group situation this experimenter was very concerned with his anxiety which made him very timid when dealing with bosses.

'Are you OK with me?'

"Yes, I'm OK with you.'

'Is there anyone here, in the group here that you feel uncomfortable with. What about so-and-so with his posh Eton and the Guards accent?'

"He's alright really, he's not a boss here."

'Thinking about your anxiety – is there some strange feeling, that you are not used to, somewhere in your body.'

…"Two feelings. One in the pit of my stomach is a lump and the other is in my throat, again a lump. Should I choose which one I'm interested in?" {d1}

'No, you can work with both although you need to choose one to start with.'

"I thought of the strange feeling in the pit of my belly first. It seems to be the most interesting."

'And the strange feeling in the pit of your belly. Do you notice some more things about it, the strange feeling in the pit of your belly.?'

"Its a soggy lump of dough." {d2}

'And a soggy lump of dough in the pit of your belly, does it have a colour, the soggy lump of dough in the pit of your belly.'

"Its grey … no its more definite than that, its orangish grey."

'And the orangish grey, soggy lump of dough in the pit of your belly, does it have a shape, this orangish grey, soggy lump of dough in the pit of your belly?'

"No, its just a roundish lump."

'And this roundish lump of orangish dough like stuff, is it very heavy this roundish lump of orangish dough like stuff or is it light in weight?'

"It is heavyish, putty like." {d3}

' And this heavyish, putty like, roundish lump of orangish

dough like stuff, is there anything else about this heavyish, putty like, roundish lump of orangish dough like stuff?'

"No, not really, its just a lump lying there."

'And this heavyish, putty like, roundish lump of orangish dough like stuff in your belly does it move about at all, this heavyish, putty like, roundish lump of orangish dough like stuff in your belly?'

"No, it doesn't move."

'And what's happened to the lump in your throat, is it still there?' {d4}

"Its still there."

'And what is it made of, the lump in your throat.'

"It is wooden, like the blocks of wood I built houses from when I was a little boy."

'And is this brick of wood particularly heavy, this brick of wood in your throat.' {d5}

"No, its just a light wood, deal I should think."

'And this light brick of wood in your throat is it coloured, this light brick of wood in your throat.'

"Its green, light green."

'And this light brick of wood in your throat, what is the size of this light brick of wood in your throat, is it a 10cm cube?'

"It is more like 20, 25 cm cube."

'And the 20 to 25 cm cube of wood in your throat is it solid or hollow?'

"It is solid."

'And this solid brick of wood in your throat, does it move about, this light brick of wood in your throat.'

"Yes, it moves up and down. Sometimes its nearer my chest and sometimes nearer my mouth." {d6}

'And the lump of putty like stuff in the pit of your belly, does it move about, this lump of putty like stuff in the pit of your belly."

"Yes it can move up towards my chest sometimes."

'And what happens if the putty like lump of stuff in the pit of your belly moves up and the wooden lump in your throat moves down and they both meet somewhere?' {d7}

"Agh! … Wow. That's peculiar, In my chest, there, somewhere was an electric shock."

'So the two bits came together. Where are they now, the two lumps that came together.'

"They aren't there any more."

'And what will happen, sometime in the future, when you meet a boss person, what will happen?'

"I'll just talk about whatever we need to talk about." {d8}

Comments on double drive. The facilitator starts by wondering what is developing. It could be a relationship problem between the experimenter and the facilitator or a group member as is typical in a gestalt group situation.

{d1} There are two felt-protofigures.

---

* The **E** and **F** abbreviations will be used throughout this chapter and, following Grove's practice, the **E**'s dialogue is designated "and"; the **F**'s 'and'. … indicates pauses.

{d2}   Stage <2>.

{d3}   Still stage <2>.

{d4}   The facilitator feels that enough is known about the lump of putty and suggests the change to the other felt-protofigure.

{d5}   This is developing information to stage <3>. N.B. The facilitator's error in referring to a brick of wood instead of a block. This was probably the facilitator's child intruding because he used to play with wooden bricks. It doesn't seem to have caused a problem.

{d6}   Movement indicates stage <3> or <4>.

{d7}   The key question for inducing movement – the experimenter might have avoided it.

{d8}   Both felt-protofigures have matured to stage <5> and solved the polarity problem.

Here two felt-protofigures appeared and each was investigated at stage <2>. The movements of each were evidently at stages <3> and <4> and the coming together as metaphors at stage <5>. Some sort of mutual annihilation occurred solving the anxiety problem.

An interesting change occurred concerning the movement of the putty lump. At stage <2> it couldn't move, at stage <4> it did move.

Black cat

This experimenter had had an affair with a bloke since school days and she was deeply in love with him, longing to be married. At 24 he left her for another girl. **E** was completely distracted with grief and lost her job.

'What do you intend for yourself?'

"I want to be able to relate to the men I meet without being bothered by memories of the jilting."

'So you want to be able to relate to men without memories of the jilting as you relate to the men you meet. What do you want instead of memories of the jilting as you relate to the men you meet?"

"I want to be able to see this man without hearing the row I had back then."

'Can you see me without hearing the row you had back then?'

"Yes indeed. I'm not interested in you in that way."

'And as you remember the row you had back then and the way it intrudes for you these days, I wonder if you are in touch with something else going on for you, within you or outside?' {b1}

After about a minute pause. "There's a black cat following me. I know that the cat's there, I don't actually see it but sometimes out of the corner of my eye I actually see it. I sometimes think I'm going a little bit potty because there isn't a cat there at all. " {b2}

'Would you like to know more about the black cat?'

"It's a nice, cuddly black cat."

'And that black cat, is it near you now? Can you sense that it is near you now?'

"Yes the cat is sitting behind this chair. Not sitting, it's cleaning itself the ways cat's do."[4] {b3}

'And the black cat, that cat, is sitting behind your chair and cleaning itself the way that cat's do and I wonder if there is anything else about the cat, sitting behind your chair, cleaning itself. Sometimes black cats have white paws and a white shirt front, is that so for that cat, which is sitting behind your chair and cleaning itself the way that cat's do'

"No, its all black. All quite black."

'And this cat sitting behind your chair, cleaning itself, is it a kitten, a medium aged cat or an old cat. How old would you think this cat could be, this cat sitting behind your chair, cleaning itself.'

"Its not a kitten, its clean and well looked after so its not a very old cat and if I go somewhere it follows me. I'm used to it really. Except that it suddenly takes me by surprise when I have not been thinking about it for a while." {b4}

'And the black cat sitting behind your chair – is there something else going on for you somewhere else in addition to the black cat sitting behind your chair.'

"Eeee, yeh, I can feel something in my stomach area." {b5}

'And there's something in your stomach area.'

"There's an empty feeling in my stomach, empty, … and I want to fill it up."

'Are you hungry?'

"Oh no, its not hunger. Its just a vague empty feeling."

'And the vague empty feeling, how large is it, the vague empty feeling?'

"About as large as a pudding in a basin, like a christmas pudding." {b6}

'And what shape is the empty space, as large as a christmas pudding?'

"Its about nine inches in diameter."

'And are you sure that there's nothing inside this empty space of about nine inches in diameter?'

"Yes, its quite empty." {b7}

'And what happens next to the empty space in your stomach, about nine inches in diameter.'

Long silent pause.

{b8} 'And the empty space of about nine inches diameter in your stomach can it move at all?'

… "No it stays quite still."

'And the black cat behind your chair, can it move at all?'

"Yes the cat moves freely." {b9}

'And the black cat behind your chair, can it move towards the empty space in your stomach?'

"The cat is curious about the empty space and is moving towards it."

'And as the black cat is moving towards the empty space in your stomach is it saying something to you, that black cat moving towards the empty space in your stomach?'

"Its purring, as it taps the empty space with its paw. Ugh, Ow, the empty space has burst." {b10}

'So what's there, where the empty space was?'

"Nothing, and the cat's gone too. Gone. … I feel disappointed really. I'm used to the cat being around.

'So you are used to the cat being around. What can happen next while the cat is not around?'

"I'll probably get a real cat. A kitten probably. Something I can feed and give attention to. We always had a cat when I grew up."

Comments on black cat. The story as told above is far from complete. In the earlier part the black cat went in the garden doing what cats do in gardens including sitting in the sun and hunting for mice and small birds. Making a much, much longer story than the one told.

{b1}   The facilitator instigates interest in a felt-protofigure.

{b2}   The black cat is the felt-protofigure – it is obviously a fantasy.

{b3}   It is interesting that the cat was behind her chair so she could not see it and yet she knows what it is doing.

{b4}   The facilitator feels stuck. Thinks: 'We're only getting an expanded description of a black cat that walks around! Yes I know what to do'.

{b5}   A second felt-protofigure.

{b6}   The facilitator is distracted by an image of a christmas pudding in a big boiler thing from his childhood.

{b7}   The facilitator is beginning to feel frustrated again.

{b8}   The facilitator feels a need to make a big suggestion.

{b9}   Stage <5>. Apotheosis!

Twitch

The **E** complained of a twitch that appeared involuntarily on the corner of his mouth, on the right hand side. {t1}

"It interrupts me when I am talking and causes spillage when I drink."

'What do you intend for yourself?'

"I want to be free of the twitch."

'What is the twitch saying to you. Can you give the twitch a voice?'

"Its saying –."

'The twitch would say – I'm saying …' {t2}

"I'm saying, take notice of me. I want you to know about me and I want other people to know, as well. But I won't say anything else. Not now." {t3}

'Can you get the twitch to twitch right now, show me how it works?' {t4}

E struggles with facial gestures for a time. 'You can act as if the twitch was occurring on the right side of your mouth. … I see you screwed up on the right side of your mouth. Do you think that that is what occurs when it is spontaneous.'

"Something like, except that the twitch occurs suddenly, and quickly."

'So can you speed up the twitch that you are acting into now?' **E** contracts a little faster. 'What happens before or after you have a twitch like this?'

"My tongue makes a kind of flip before the twitch occurs." {t5}

'And your tongue flips before the twitch occurs and what happens before the tongue flips before the twitch occurs?'

"I want to say something."

'And what do you want to say before your tongue flips and your lip twitches, what do you want to say?'

"I don't know. I don't remember."

'And what don't you know and what don't you remember

before your tongue flips and your lip twitches.' {t6}

'I'm feeling very confused. There is a sort of weight on my tongue.' {t7}

'And there is a weight on your tongue, how heavy is the weight on your tongue.'

"About as much as a pat of butter. Perhaps half a pat."

'And the weight that is as much as half a pat of butter, what size is the weight that is as much as half a pat of butter.

"Its about as big as a match box. Yes flat like a match box as it lies on my tongue." {t8}

'And the weight that is as much as half a pat of butter and is match box sized lying on your tongue, is it empty or full of something, the weight that is as much as half a pat of butter and is match box sized lying on your tongue.'

"It contains explosive." {t9}

'So the weight that is as much as half a pat of butter and is match box sized lying on your tongue contains explosive. And what happens next?'

"I put the match box containing explosive in a bucket of sand and melt the explosive with a jet of steam."

'So you put the match box and the explosive in a bucket of sand and melt the explosive with a jet of steam. What happens next?'

"The explosive has melted into the sand and disappeared. I feel very relieved that the explosive has been dealt with and the danger gone."

'And what happens next for you as you have dealt with the explosive so that the danger is gone?' {t10}

"I feel very relieved, no tension now, and I'm pleased with this outcome." {t11}

The facilitator heard from the experimenter after three months or so. The facial tic had not recurred and the experimenter was finding it easier to hold conversations.

Comments on twitch. The experimenter evidently knew his felt-protofigure very well and was evidently desperate to say something devastating and stopped himself . By using the device of disposal of explosive he was able metaphorically to remove his impediment.

{t1}   The twitch is the felt-protofigure.

{t2}   The experimenter is encouraged to own what his twitch says.

{t3}   The facilitator respects this request and changes the subject.

{t4}   Suggesting enactment. The activity is a gestalt technique which helps the experimenter to know his felt-protofigure.

{t5}   Involvement and movement of the tongue at stage <3>.

{t6}   The experimenter knows that he does not remember but the facilitator knows that there is a metaphorical protofigure there, somewhere.

{t7}   Confusion leads to a new felt-protofigure, the weight on his tongue.

{t8}   It is as big as a match box and not a match box. It is not a noun yet.

{t9}   There is the noun, explosive stuff at stage <4> –

{t10}   – which he promptly disposes of at stage <5>.

{t11}   The experimenter is pleased, etc., at stage <6> and the end of the episode.

Car crash

This experimenter (**E**) was a passenger in a car crash that killed her father. She was then 6 years old and is now 32. She sat with the facilitator (**F**) and complained of flash backs, sometimes during the day and as nightmares when she woke up feeling terrified.

[**F** notices that **E** looks worried and is not engaged with **F**. She is probably remembering flash back incidents].

'As you remember the accident are you aware of anything else happening inside you.' {c1}

"I'm rather excited and wondering what you want me to do."

'And as you quieten down to a meditation state what would you like to find for yourself as you quieten down to a meditation state? {c2}

…"I have an odd ache in my stomach." {c3}

'And an odd ache in your stomach. Would you like to know more about the odd ache in your stomach?

"OK."

'The odd ache in your stomach, what shape is the odd ache in your stomach?' {c4}

"Its like a large ball" {c5}

'And an odd ache like a large ball. How large is the odd ache like a large ball in your stomach.'

"As large as a football."

'So the odd ache in your stomach is as large as a football. Is it coloured, the odd ache in your stomach that is as large as a football?'

"Not coloured. Its rolling around." {c6}

'And the ball as large as a football is rolling around in your stomach. What would the ball like to do as it is rolling around in your stomach?'

"It wants to escape." {c7}

'And where would the ball escape to?' {c8}

…"Over the hills and far away." {c9}

'And the ball rolling in your stomach wants to escape over the hills and far away. What is stopping the ball as it wants to escape –'

"Its gone." {c10}

'And the ball rolling in your stomach has gone. Where has it gone, the ball that was rolling in your stomach?'

"Its in the kitchen. My little boy is kicking it around. He is very happy but I'm afraid he will break something." {c11}.

[**F** notices that **E** is more relaxed and smiling as she talks about her boy].

'And your little boy is happily kicking the ball around in your kitchen. What happens next as your little boy is happily kicking the ball around in your kitchen?

"Nothing much. I will be making his supper soon." {c12}

'So how will you sleep tonight?'

"OK, I expect. I'll let you know."

Comments on Car Crash. Follow up after weeks or months usually showed that she would never have flash backs again.

{c1}   **F** is inviting **E** to become aware of subliminal impressions.

{c2}   After suggesting that **E** become quieter **F** repeats his invitation in a different grammatical form.

{c3}   **E** felt at stage <1> and then talked about, stage <2>

a felt-protofigure, the ache in the stomach.

{c4}   **F** begins to ask questions aimed at filling out details about the felt-protofigure

{c5}   **E** is fantasizing about the ball. More information for stage <2>.

{c6}   **E**'s ball is moving so it is more solid, noun like and probably at stage <3>.

{c7}   Escape! A dramatic change, stage <5>

{c8}   **F** invites an increase in available options.

{c9}   **E** indulges in poetic fantasy [this often occurs as an **E** is remembering child or youth events].

{c10} The escape has occurred. **E** has freed herself from the "odd ache in [her] stomach"

{c11} **E** is no longer concerned about fantasy balls and identifies it with her son's ball. She loves him and is happy.

{c12} **E** is not concerned about flash backs.

This has been, as emphasized above, a created story based on reality. In reality the experimenter develops information from his or her protofigure in the way described except that interruptions may occur. For example **E** may suddenly become interested in a new felt-protofigure, perhaps at {c9} in the above sequence. The process of developing information at stage <2> then begins again.

It is interesting that **E** identified her stomach symptom with a ball. Grove[3] suggested that we could account for that if we knew more about the circumstances of the car accident. In this case there might have been a brother's football rolling around in the car as it crashed and that that football had been centre for the 6 years old's attention at the time.

Vague volatility

The experimenter presented in a group, complaining of constant fatigue and "lack of interest in anything in particular." She said she was not depressed because sometimes she became very energetic, happily dancing around until she "flopped out."

'So … what do you intend for yourself?'

"I want to be more ordinary. People at work complain that they don't like me when I'm morose."

'And as you talk about being morose, what happens next or what happened before you were morose.' {v1}

… "Well … I think a lot."

'And as you think a lot are you aware of a strange feeling, or something somewhere in your body?' {v2}

"Yes, but not in me. There's a cloud hanging over me." {v3}

'And what is the cloud like, the cloud that is hanging over you?' {v4}.

"It's big, towering and black."

'And a big black, towering cloud is hanging over you. What would you like to have happen to the big black, towering cloud that is hanging over you?'

"I want it to go away." {v5}

'And how can the big black, towering cloud hanging over you go away?'

"A wind can come and blow it away." {v6}

'And how can the wind come and blow away the big black,

towering cloud that is hanging over you?'

"It has to be really bad weather."

'So when will the weather be bad enough to blow away the big black, towering cloud hanging over you?'

"In the winter time. The weather is rough in November."

'And you are going to wait until November for the weather to be bad enough to blow away the big black, towering cloud hanging over you.'

"No, well, a summer storm will come, perhaps. But I don't want a storm. I want peace and quiet and space to think about things." {v7}

'And the big black, towering cloud hanging over you. What has happened to the big black, towering cloud hanging over you?'

"Its nice now. A sunset, with lots of red and orange, warm colours. I feel more alert as I watch the sun set." {v8}

'And as the sun sets, are you sure that the sun is setting and not rising in the display of red and orange clouds.' {v9}

"Oh yes. It's a sun set. The sun is going away for the night. It will be a new day tomorrow."

'And what will your colleagues say to you tomorrow after the sun has risen?'

"I don't much care really. It will be all right, I suppose, unless I dance around." {v10}

Comments on vague volatility. David Grove[3] always saw patients 1:1 when I was with him. I have found benefit in using his approach in a group situation because bystanders get the ideas and want to have a go themselves. Also sessions seem to take less time in group than 1:1; the experimenter seems more inclined to concentrate attention.

{v1}   Ericson style trance inducer; 'after' versus 'before'.

{v2}   **F** invites attention to a possible felt-protofigure.

{v3}   The felt-protofigure is outside the body boundaries.

{v4}   **F** encourages **E** to obtain more information about the felt-protofigure at gestalt figure stage <2>.
        'What is the cloud like' is equivalent to asking: 'Would you like to know more about that cloud?'

{v5}   **E** wants the cloud to go away, movement indicates stage <3>.

{v6}   **E** begins to explore options; what can happen to the black cloud.

{v7}   **E** makes energetic changes at stage <4>

{v8}   **E** makes a total change; black clouds become a colourful sunset at stage <5>.

{v9}   **F** challenges **E** about the ending.

{v0}   **E** is preoccupied with something else and evidently satisfied <6>.

The black cloud felt-protofigure stayed outside the body as it morphed to make a sunset. Sometimes the change involves something inside the body.

General comments

I hypothesize that, even if the experimenter does not describe a trauma, there would have been a trauma that is now forgotten. This could be so for the experimenter, described above, who had trouble with the right hand join of his lips. The facilitator may be very curious about antecedents but has best keep his curiosity to himself. Sufficit that the therapy session is followed by contentment for the experimenter.

When choosing examples for this account I thought it would be most redolent if I started with a butterfly. However lumps present more often so I chose a pair of them. I thought it interesting that, for the Double Driver, one lump was soggy and the other hard, typical polar opposites. Resolution of these was rather like resolution of an impasse but was explosive for the experimenter.

A more realistic fantasy explosion ocurred for the Twitcher in an event that most have been related to his life experience. It would have been interference if the facilitator had diverted interest to his curiosity about it.

The black cat lady had one external and one internal protofigure, a cat an an empty space, a living creature and a very dead void; again kind of opposites.

When only one protofigure is evident it can be hypothesized that there is a second one that has not made itself apparant. This does not seem to matter as resolution is available anyway.

The vaguely volatile man had one external protofigure which he resolved outside himself, like so many finalès, going into the sunset.

When two protofigures met a shock and startle reaction occurred, which was described by the experimenter as an electric discharge or an explosion.

Notes and references

1   Gendlin, 1996.

2   Grove, D, 1989. I had the pleasure of joining many of David Grove training sessions where he always helped experimenters with traumas.

3   Kopp, 1995.

—=(☆)=—

Chapter 10

Gestalt Protofigure Analysis

Before a gestalt figure appears
lying deep in a vortex of nerves
is a protofigure.

The procedure for carrying out protofigure therapy is a skeleton on which the facilitator hangs robes of clean language. In this chapter the gaudy robes are here but they will only be given major attention in Appendix 1.

During my early interest in protofigures I was aware of natural phenomena, the sense-data, lumps, butterflies, clouds with silver linings, etc., data which developed as I gave it attention and found that it spontaneously developed along the track that was already well known to me as the stages of development of a gestalt figure. (Chapter 2 and the Frontispiece).

In the following exposition the early observational approach will be avoided, the stages of development of a gestalt figure will be accepted as a justified hypotheses and the protofigure development procedure will be set out on those lines. So the first observation is to know of the existence of the felt-sense, the precursor of a new gestalt figure, the protofigure*.

---

* In this description the protofigures will be referred to as a gestalt protofigure and abreviated to GPF1, GPF2, etc. In reality the name of the felt-sense is used; dog, log, fog, etc. The nature of each GPF changes as it goes through the gestalt figure stages.

On presentation

The first task for the facilitator is to encourage the experimenter to talk about his or her felt-protofigure and to learn more about it using clean language (Appendix 1) and ideas from Chapter 2 –

what is it, colour, size, shape, texture, etc?	stage <2>
where is it, belly, chest, throat, outside, etc?	stage <2>
is it moving and or changing?	→ stage <3>
are there alternatives, options?	→ stage <4>

Therapeutic work with felt-protofigures may start from one of several positions. If the experimenter arrives and says "I've got a tic in my right eye" the tic is the felt-protofigure.

Otherwise the experimenter may talk vaguely about recurring dreams or day time fantasies or describe recurring events that disconcert him or her. The latter may include statements like –

- "I get afraid every time I see a dog. I don't know why – I don't remember having trouble with a dog."
- "Whenever I prepare to leave home I get butterflies in my stomach."
- "I feel so bored and tired all the time."
- "I become tongue tied in the presence of beautiful women."
- "I had this nasty car accident a few years ago and now have memory flash backs at awkward moments

and nightmares."

- "We had a fire in the kitchen years ago and I ran screaming from the house. Now, whenever someone lights a match or a bright light dazzles me I tremble all over and my blood runs cold."
- "I catch myself grinding my teeth. It seems to happen when I talk to a senior man."
- "It's odd. When I enter a bank I feel as if there was a wall in front of me."
- "I'm afraid to go out in the dark. I expect a strange man to attack me. I see the flash of his dagger and I feel queasy."

The facilitator encourages the experimenter to divert attention from these events and to concentrate on the knowledge he or she has of the felt-protofigure which may otherwise appear to be a side issue.

The facilitator may see signs of discomfort in the experimenter. For example, wringing of hands and general fidgeting indicate anxiety, as do rapid eye scan movements, as if the experimenter was expecting something alarming that probably originally occurred in childhood or adolescence. The facilitator may decide to share his or her observation with the experimenter but has the general intention of avoiding bringing on an episode of debilitating affect. He or she also intends to use only Grove's clean language[1] with a very gentle, supporting, non-intrusive, soft approach – as will be explained later.

In general, the facilitator says something like "while you are aware of your interest in the trauma (name it!) do you also have something different going on somewhere?" This would be the protofigure.

Forestalling traumatic affect

The facilitator may guess that the experimenter is, while talking about the occurrence of a trauma, beginning to feel the catastrophic affect of the trauma. If affect develops it would be necessary to postpone the therapy session. It is up to the facilitator to use his experience to forestall this event. To a young person he may say something like; 'Cool it, man!' To an older person; 'And you could give your attention to your peaceful and quiet self'. 'And rather than get hot under the collar, you could give your attention to your peaceful and quiet self.' Meanwhile the facilitator models placidity and is not aroused emotionally or intellectually, in other words, does not become embroiled in what might be going on for the experimenter.

How many protofigures occur?

As indicated in Chapter 9 the experimenter may produce one or two protofigures. Development of one protofigure is straight forward but the appearance of aspects of two ptrotofigures may be rather muddled as the experimenter switches attention from one to the other. The account to be presented concerns two protofigures but, for clarity, begins as if for only one.

Recognized felt-protofigure seeds
These are all experimenters expressions.

Seeds	*where found*
Internal	
wooliness	in head
throbbing	head, chest
ache	head, stomach[2]
turd	in head
tinitus	ears
motor tic	one or both eyes
tears	one or both eyes
pulse	on fore head
deadness	face
blushing	face
stiff muscle	upper lip
muscular twitch	edge of mouth
frog	throat
heavy weights	on shoulders
palpitations	heart
butterflies	stomach
lump	stomach, chest, throat or mouth
ball	in stomach
knot	in stomach, in head
wound up clock spring	in stomach
gutted	in stomach
churning	in stomach
sinking feeling	in stomach
fluttering	in stomach, on hand
itching	on belly and forearms
sleepy	body
cold sweat	body
cold shivers	back
my blood runs cold	body
spaced out	general body
wrench	inside me
heaviness	hand
heaviness	feet
heaviness	heart
External	
wall	in front of bank
closed door	bedroom
river	blocking a road
mill pond	country side
black clouds	hanging over
everything is up in the air	

```
┌────────────────────────────────────────┐
│          Possible characteristics        │
│      that the protofigure may develop     │
│                                           │
│  phase        solid, liquid, gas          │
│  geometry     box, sphere, irregular mass │
│               hollow or solid             │
│  colour       black, red, etc.            │
│  texture                                  │
│  of surface   rough, glassy, smooth       │
│  identity     foot, tennis, snooker, etc., ball │
│               apple, orange, carrot       │
│  external     clouds, walls, hedges       │
└────────────────────────────────────────┘
```

The felt-protofigure; stage <2>

There are four ways in which protofigures may be *felt* –
- An experimenter may be aware, without prompting, of an odd phenomenon somewhere in his or her body. Examples are motor tics and various psychosomatic symptoms.
- A felt-protofigure may come into awareness during special circumstances. *Example*: On approaching a rostrum to give a speech to a multitude, the speaker becomes aware of "butterflies in my stomach."
- During everyday events a flash-back from a trauma dominates awareness. There may be no obvious cause for this. Likewise a nightmare interrupts sleep.
- A protofigure may be called into awareness by a facilitator under special circumstances. *Example*: The experimenter is talking about a nightmare that occurred last night. The facilitator suggests a quietening down to a meditation state and then a search for 'something unusual somewhere in the body or maybe outside' The experimenter then says something like: "Yes. I have a throbbing pain in my lower back." This is the protofigure. Examples of other felt-protofigures which have been used as seeds for development are set out in the box on page 150.

In the normal course of events attention is not given to the protofigure – a speaker avoids attention to butterflies in his belly and gets on with his job, speaking. However, it is not usually possible to avoid the dominance of a flash back. The following remarks will concentrate on the likes of fantasy insects, nightmares, etc., and what can be done with them.

Grove classified what I now call protofigures as –
- Internal: e.g., "A rock behind my eyes."
- External: e.g., "Fog swirling round my feet."

This suited him as he worked with metaphors. I have not found it necessary to treat the two situations differently, because the external protofigure usually becomes internalized as the development of figure stages proceeds.

Having discovered a protofigure the next step is to increase knowledge about it – some observations are detailed in the "Possible Characteristics" box. Eventually the protofigure is fully characterised and the need to move on is felt.

Clean language is the facilitatory tool (Appendix 1). By using it the facilitator focuses the experimenters attention, producing a state "of concentrated self-absorption" as the experimenter is interested more and more in his own discoveries. The facilitator does not instruct or intrude in the experimenters process. There is no formal induction of a trance state which occurs anyway, as it does with Milton Erickson's procedures. More aspects of clean language are –

- The facilitator's first key question is very important in that it must not orientate the experimenter in any particular mode of operation and language, past, auditory, metaphorical, etc. The experimenter sets his or her own scene. Grove lists the following recommended openers for the facilitator; 'What is it that you want?' 'What do you need to have happen?' 'What would be helpful?' 'What would you like?' I find it more productive to ask –

'What do you intend to
have for yourself?'

Or something like:

'You can wonder
what you intend
to have happen?'

Moving on from stage <2> to <3>

Stage <3> is characterised by recognizing energy in the system so that movement and changes occur. This may occur spontaneously or can be promoted by the facilitator with questions involving use of the phrases like – "What happens next?" "Does the GPF want to move or change?"

Moving on from stage <3> to <4>

The characteristic property of things at stage <4> is that options become available and it may be necessary for the experimenter to make choices among them. Some useful promotional questions from the facilitator are in the style of – 'Does GPF1 want to go to wherever or does GPF2 want to go too.' 'What happens next when so-and-so happens?'

The meeting of stage <5>s

Eventually the facilitator senses that the development of both protofigures is complete and he checks –
'Is GPF2 moving about?' and experimenter agrees or disagrees.
'Is GPF1 moving about?' and experimenter agrees or disagrees.
Or the other way round, with GPF1 first. Then 'What happens if PFT1 moves towards GPF2 and GPF2 moves towards GPF1?'

The experimenter feels something dramatic happen in his or her body and may report explosive, electric spark sensations, a flash of heat or chill or something similar.

Stages <6> to <8>

After the drama the experimenter says he or she feels finished. The facilitator notices a marked change in the disposition of the experimenter. Expressions of satisfaction occur spontaneously or can be evoked by the facilitator.

Future projection

The purpose of this adventure was therapy and it is possible to check that this has in fact happened. This will happen as time passes as the nervous tic does not recur, nightmares and daytime flash backs do not recur and other of the previously unwanted signs and symptoms do not recur.

However that takes a passage of time, probably on a week by week scale. It is possible to obtain an immediate indication by using a future projection method. The experimenter becomes as calm as possible, gets into a trancey meditation state and then imagines a time in the future when the signs or symptoms would be expected. The therapy is successful, as described above, if the events of the trauma or problem are remembered without the expected affect, i.e., a daytime flash back does not occur and morning awakening is accompanied by realisation that there was no nightmare in the night.

Jumble

It is not possible to predict precisely what will happen during a protofigure therapy session. This is largely because there are two protofigures involved and it is not possible to predict when the development of the first will be interrupted by birth of the second.

It is also possible that the session will produce only one protofigure in which case an alternative ending will occur.

The single protofigure event

The facilitator supports the experimenter's work on a protofigure while expecting the second at any time including simultaneously. But this may not occur.

The single protofigure is developed as far as stage <5> when it will be found to be spontaneously transformed and it is then clear that the therapy is complete and can be checked using future projection as described above.

Discussion

Regression

The experimeter starts a session as an adult and may talk about being worried, then, due course of time, becomes child like and says: "I'm scared!" There is a chance, then, that the experimenter is in a regressed state, probably very young, so it is advantageous to the experimenter if all the facilitators statements and questions can be simplified and manageable by a child. Many traumas occur in the early years although regression to teenage years often occurs when conflict with adults at home was most intense.

Metaphors

The descriptions of actions, etc., at the various figure stages are, as David Grove surmised, metaphors. These were intuitively developed by him to reach curative positions but with nothing like the regularity, smoothness and relevance provided by thinking of them as gestalt figure stages. Lawley and Tompkins[3] also consider the observed phenomena as metaphors.

David Grove says that increasing knowledge about a felt-sense leads to a conclusion that it is a noun (he says it can be drawn). My experience supports his idea. The major difference between David and I is that he talks about metaphors, for which he has no method for promoting maturation, and I talk about gestalt figure stages with clear maturation in mind, stage by stage.

It also might be useful, though, sometimes, to think of the new gestalt stages in terms of them being metaphors.

Intentions intended

It was, as mentioned above, a general principle, often repeated by David Grove[1], that the most effective approach to the experimenter occurred if one presumed he or she was a child. This does not necessarily occur if the primary leading question is; "What do you want for yourself" because the word "want" has two separate and confusing meanings – adults sort these out because they are accustomed to the confusion. "Want" and derivative words are associated with "needs" and also with "deficiency, lack, poverty, privation, longing for" (COED) and lots more. To avoid this ambiguity I use "intend" and derived words, providing a clear focus to aid the experimenter's concentration and decision about purposes.

Self therapy

Anyone who thoroughly understands the protofigure therapy system can settle comfortably in a meditation state to know about a problem (without affect), guide self to find a felt-protofigure and proceed from there. Success follows steady concentration in the here and now.

It is particularly important that, in the facilitator role, only clean language is used.

The facilitator ↔ experimenter interaction must be set up very clearly so that the practitioner knows when he or she is in one role and not the other and then exchanges roles whilst maintaining clarity. I have found the following procedures to be very effective –

• Use two chairs. Change roles by changing chairs.

• A way that does not involve such large movements uses two hands. Sit with both hands in the lap. Raise one hand when talking as the experimenter (I use my left hand) and raise the other hand when being the facilitator. The hands can gesture at one another.

• Verbal gymnastics. As facilitator I say something like: 'And as facilitator I ask you, experimenter, what shape is the lump in your stomach.' The experimenter might reply something like: "The lump is cubical, transparent as glass and the

The relationship of <u>felt-protofigures</u> to gestalt phenomena.

Generalized characteristics of the gestalt figure stages

Aspects of person to person engagement are italicized.

<0> Something stirs within the mind /brain, a protofigure. Its nature is partly determined by previous experiences with gestalt figures and partly by stimuli from the environment.

<1> no verbalization by thought or voice. Vague feelings that something is about to happen, that something is missing or required. A felt-protofigure.

<2> awareness, verbalization, thoughts and affect particularly about a problem. Expression of the need for nurture. Knowing intentions, interests, concerns, curiosities and desires. *A partner is approached – hello!*

<3> rising energy, excitement and interest and enthusiasm for satisfying needs. *A partner is contacted and asked – how are you?*

<4> mobilization of resources, choice among options, contact at the beginning of action. *Internal decisions are made about things to talk about and do with partner.*

<5> engagement with the need and hunger satisfying action, curiosities are satisfied. Consummating action; apotheosis, apogee. The activities of self-nurturing are proceeding. *A full conversation develops with the partner.*

<6> satisfaction of needs, wants, hungers, curiosities are complete and engagement with the necessary activities ends. *Engagement with the partner ceases – caio!*

<7> Awareness of the ending; feelings of satisfaction, pleasure, fulfilment, satiation and gratification. The endeavour is rewarded by warm, happy feelings and one gives thanks to self for a job well done. *Wander away from partner feeling warm, happy and good inside.*

<8> Withdrawal ready to deal with a new gestalt figure with a pleasant afterglow feeling.

<0'> **Grounded**. The events of this series of gestalt figures are remembered and can form the basis for future gestalt figures.

Aspects of behaviour to seek when encouraging completion of a sequence of gestalt figure stages.

<0> Expect that the mind/brain will have adequate resources and gestalt protofigures.

<1> In a meditation state contact a felt-protofigure.

<2> Foster attention and awareness of thoughts and affect. Expect an interest in satisfaction of needs, interests, concerns and desires.

<3> Expect and encourage increasing energy and excitement.

<4> Expect an interest in choosing among options, including resources and interest in how to begin contact and engagement activity.

<5> Expect enthusiastic activity in satisfying needs, wants, hungers and curiosities.

<6> Expect and encourage appreciation of the activity just completed.

<7> Expect awareness of completion with attendant beneficial feelings of satisfaction.

<8> Expect a clear sense of withdrawal from what has been going on and an expectation of enjoyment of the next gestalt figure to arise.

<0'> Expect a positive attitude towards encouragement of memories of what has just happened.

Neurology

Major neuron connections relevant to traumas

Hypotheses of LeDoux[5], Papez[6], Green and Ostrander[7].

What we commonly call the mind is a set of operations carried out by the brain … all behaviour is the result of brain function … and behavioural disorders … disorders of affect and cognition … are disturbances of brain function.

Kandel and collegues[4].

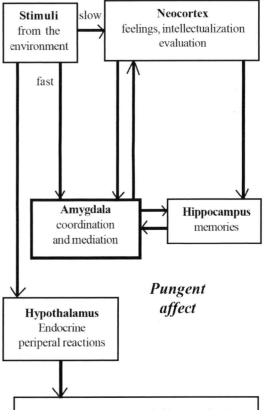

When prompted the implicit memory [*ibid,* p 1247] of a panic disorder trauma is recallable complete with catastrophic affect because the neocortical aspect is connected to the affect centre, the amygdala. For example, panic disorder. (*ibid,* p 1221) If, as a hypothesis, the neocortical aspect can be recalled without amygdalal involvement the remembered event is without catastrophic affect, much to the relief of the person concerned. The unused neuron connection, neocortex to amygdala, becomes resorbed [*ibid,* p 1952] while new neurons are generated (*ibid,* p 1275) if the person feels pleasure at the relief.

The second hypothesis is that a nub of neurons of the trauma lurk near the conscious / non-conscious boundary and cross to become conscious as felt-senses which can be developed to form, as was originally conceived by Grove, metaphors for the neocortical state. These metaphors are now seen in a gestalt context as felt-protofigures which may occur spontaneously for some people who effectively heal themselves.

The third hypothesis is that the use of clean language as described by Grove[1] facilitates development of the felt-protofigures to form stages of a new gestalt figure. Grove's soft and slow approach fits with the neurological conclusion [Kandel, *et el,*[4]] that a slow repetitive approach is most effective for learning.

The fourth hypothesis is that an alterred metaphor can be remembered and form a neuronal nucleus that can function instead of the original traumatic nucleus so that the client can remember the traumatic incident without debilitating affect and never be troubled by it again.

These four hypotheses are evidently satisfactory explanations of the processes of felt-protofigure therapy, because Grove's anti-trauma therapy is very effective, especially in comparison with any form of preceding therapeutic processes.

David Grove did not discuss these hypotheses[3] as such. He used his processes very effectively, noticing the decommissioning of affect and avoided theorizing.

Corollary: It is natural to place emphasis in this account on the occurrence of major traumas. The procedure described for use with major traumas works just as effectively with minor traumas and simpler problems.

Further experience may indicate value for this felt-protofigure procedure for stress conditions, like post traumatic stress disorder (PTSD).

Gestalt therapy has thus been provided with an extended neurological base. Instead of vaguely discussing ground processes we are provided with sound neurophysiological processes, protofigures.

shape of a die." And the facilitator might say: 'And the lump in your stomach is cubical, transparent as glass and the shape of a die – does it have the numbers on it?' The the experimenter might say: "No. Its just the cubical shape and size of a die." The facilitator might go on: 'And the lump is cubical, transparent as glass and the shape of a die. What colour is the lump which is cubical, transparent as glass and the shape of a die.' Such verbal gymnastics are difficult to maintain.

And so he or she can go on, pretending to be two people and exploring the ramifications of a problem with an outcome similar to that of an interaction between two people.

Nomenclature

My attention was early focussed on protofigures because they are immediately related to the felt-senses, rocks in the throat, etc.

I could have called my new approach to gestalt therapy, impasse therapy since intervention by the facilitator always generated the impasse; it was no longer a matter of serendipity.

On the other hand, detection of the felt-protofigure was an act of the experimenter alone, with gentle encouragement from the facilitator. This was a form of self therapy.

Of these three terms protofigure therapy, auto-impasse therapy and self therapy I preferred the first because it kept the nomenclature within the gestalt orbit.

Coertion

My general intention is to have the experimenter functioning freely so I need to consider if the approaches used in protofigure therapy are unacceptably coercive.

Grove's clean language is selected as being the least coercive available. 'What happened, what is happening, what might happen' are in the flow of everyday life where something is happening even if it is sleep.

'What would lump like to do next?' is a use of fantasy, adapted from Grove, that I can't justify but which works as an aid to the experimenter.

In my description the facilitator seeks and promotes development of gestalt figures. These are a natural events, if the gestalt hypothesis is true, because the experimenter will have spent his life developing one gestalt after another.

Health

If we are to know what unhealthy people are like in their dissociated states it is necessary to know what healthy people are like in their associated states. (Appendix 2, page 167). However, we must be sure to find real healthy people. Actors mocking health on TV are not suitable subjects for contemplation. Real people are out in the world, in the street, in super markets and on the beach.

A psychologically healthy person may search for felt-protofigures of deleterious origin and be unsuccessful. He or she can take pleasure in that. However if there are in the past the opposite states to those of trauma, states of great joy, these may produce felt-protofigures leading to discovery of more joy and euphoria.

Gestalt

Hypothesis: a person who has an
overt problem also has a covert
solution to the problem
that the person is rarely able to reach
without external help.

Protofigure therapy as described in this paper differs from the general run of gestalt therapy as it originated with Fritz and Laura Perls in being more didactic and more guided by processes. Most gestalt therapists intervene in the processes of experimenters to the least possible extent and to the point of passivity. The current essay is more assertive but certainly not as aggressive as Fritz Perls was described as being. The emphasis remains solidly with doing therapy in the particular time and place of the work.

Traditional gestalt practice was experimental in the way emulated here and placed emphasis on the exploratory value of role play though not in the exacting manner described here where an aspect of personality, the new protofigure, is thoroughly developed.

Traditional gestalt practice often inadvertently produced a state of semi-trance, locked in a conflict of opposing trends, called the impasse. The fusion of two protofigures at climax seems to me to produce a state that is very similar to that of the impasse. I hypothesize that the protofigure method has induced a state of impasse.

Notes and references

1   Grove and Panzer, (1989) and personal communications.
2   Beware. A sensation in the belly may be due to neither fantasy butterfly nor caterpillar but be an incipient ulcer – a job for a physician or surgeon.
3   Lawley and Tompkins, 2000.
4   Kandel, Schwartz & Jessel, 2000.
5   LeDoux, in Kandel et al[3], p 986.
6   Papez, in Kandel et al[3], p 987.
7   Green and Ostrander, 2009.

The kind of story once told to small children

Once upon a time a youthful knight wandered, as youthful knights did in those days, through a dense forest. From a high hill he saw in the distance, a tower. So he approached this stone eminence wondering if it pertained anything significant.

Then he saw, swirling from a very high window, a great bundle of hair. He knew at once what he had to do. He gathered the hair into a skein and climbed up it. As he did so he heard a loud woman's voice wailing from above.

At the window sill he heaved himself into a darkened room and stood aghast as a beautiful youthful princess sang abuse at him.

"You've nearly torn my hair out from its roots!
You should have entered the door
and ascended the stairs."

—=(☆)=—

Coda

The Contemplative and Experimental Life

Curiosity

My most real self is active when
my figures emerge and retreat
with respect to my ground,
spontaneously, with life,
elasticity, fluidity
and passion.

This is the point where I decide to finish my presentation by making summaries and evaluations.

At the outset of making this book I set myself the task of making a presentation in the context of the key concepts of gestalt psychology and psychotherapy. The result is a story that began with the work of scientists who, at the begining of the 1900s, recognized the characteristics of phenomena that take the adjective, gestalt.

Science

As I take a broad view I am aware of a stream of hypotheses tha t seem to be a true representation of natural phenomena. They will only graduate from hypotheses to real science when other people have replicated the work and come to similar conclusions. Meanwhile I summarize my observations as recognition of the value of –

- The attitude of a scientist to gestalt theory and practice – particularly to Wertheimer's account of gestalt theory, to the dialectic of polarity phenomena and to mantaining a dynamic attitude to change.
- Analysing failure of gestalt development in terms of interruption of gestalt figure stage circuits and their evolution.
- The grounding of the stages of the gestalt figure circuit – particularly in depression.
- understanding healthy processes and distinguishing them from the processes of lack of health and the relationship of these to associative and dissociative processes.
- differentiating clearly between the gestalt processes of perception, needs and action.
- using "clean language" in establishing an empathetic relationship and in helping an experimenter to more clearly understand his or her predicament.
- of ideas about interruption processes in distinguishing logical levels; primary figure stage analysis, secondary recognition of the inter-personal engagement functions of projection, introjection, etc., and the tertiary effects of shame, guilt, etc.
- recognizing that physiological problems, respiratory, metabolic, neurological, etc., may disrupt behaviour and personality manifestations.
- Darwinian evolutionary ideas in drawing

attention to the varied effects on behaviour of inherited genetic versus social factors.

- the use, in dealing with traumas, of recognition of protofigures in the form of lumps in the throat, clouds hanging over, etc., and developing then to form full gestalt figures.

Evolving experimenters

The purpose of the experiment
is to help the experimenter
find out for himself
how he interrupts himself
and prevents himself from succeeding.

F. Perls[1]

In my experience experimenters working with David Grove have had success in dealing with traumas, becoming able to live without haunting recurrences of catastrophic affect. This induced me to wonder if the protofigure method (Chapter 9) could be used for much less dramatic events. Preliminary experiences have indicated that the method has potential for benefiting people with problems of lesser dramatic impact than traumas. A storry entiled Wooden Worries is told in a box near here.

Another story concerning catastrophic expectations aflicted a student who was worried about exams and the dark clouds that hung over him. Given attention the dark clouds changed into a dark forrest with a lake where the sun shone as he floated in a rowing boat.

I hope to report more experiences in B.G.J. in due course of time.

Anticipation that further innovations and developments of the characteristics of protofigures may be of value in helping experimenters with problems other than those associated with traumas. These may include –

- Interpersonal relationship problems.
- Problems associated with failure of cooperation with other people.
- Problems of being shy of altruistic behaviours.
- Exploitation of gestalt protofigures in the fostering of behaviours beneficial for the experimenter.
- Scientific gestalt in the context of the interaction of groups of more than two people at a time.

I am imagining that
imposing a scientific attitude
dephlogisticates my gestalt attitude.
Yet little do I know
what the future holds
as the dross departs
and the glowing treasure
flows away out of my control.
Verses are not necessarily significantly
different or the same as poetry.

Scientific Gestalt

Tell me about it and I will forget
Demonstrate it to me and I may forget
Have me do it and I will remember.

Iroquois wisdom

The account in the above chapters substantiates a concept of a form of gestalt therapy based on firm scientific principles. In the first place deductions are made from observations. Secondly it could be possible to invalidate the deductions as described by Popper. The main features of the claim to scientific validity are –

- The occurrence of the Cleveland circuit of stages is certainly validated by the many facilitators who rely on them.
- A healthy gestalt is not holy – only a disrupted, incomplete gestalt has a hole in it (page 1) though this may not be apparent if it is blocked by introjected rubbish.
- The phenomena of felt-protofigures and of the development of them to make fully grown gestalt figures is based on the observations of Gendlin and Grove. This is validation enough for me to establish scientific nature for them.
- The hypothesis that depressed people are unable to ground their figure stages has yet to be validated by other observers.
- The observation concerning errors of engagement, from projection to shame, are more literary in quality but attach seamlessly to the scientific concepts.

Gestalt

History again

Anyone who can formulate
an intention in clean language
can realize that intention.

As my Introduction was my head so my Coda is my tail. Here I assess what has happened to me since first preparing the Introduction over twentyfive years ago. As explained in the Introduction I set out to write a book to summarize and tell myself about my knowledge about gestalt psychodynamic analysis. The book has changed as the knowledge and experience have developed.

During all this writing activity I have continued to be involved in the various aspects of community life in my Village, Charmouth, Dorset and indeed nearly got myself on the Civil Parish Council – I lost by one vote. I play music for the local Morris dance side – being unable to cavort myself.

Wooden worries

Justine came fretting over a "wandering husband" and felt unable to do anything about it. She suspected infidelity. In therapy she reported frequent dreams in which she tripped over a step and fell.

While meditating on these dreams she was invited to notice anything else going on for her and she reported a "thumping sensation in her forehead". On developing this protofigure it turned into a powerful locomotive which eventually "sped of into the night". A week later she reported challenging her husband who produced evidence of earning money on the side by doing odd carpentary jobs. She was happy with his explanation though not happy with his reticence.

Her initial predicament could not be classified as traumatic; it was only worry and catastrophic expectations.

To foster my writing activities I gave up production work for the **British Gestalt Journal**: I still support it and look forward to it being less of an existentialist broadsheet and more of a Gestalt Journal. I also gave up my weekly commutation to London to run a therapy group.

I was afraid of senility and I am still concerned.

This book has been my essay into integrating all that I know about my selves as they are within my hypothetical psychological interface.

Has all this discussion about types of figures and stages of development supported the initial propositions set out in the Introduction or have the complications concerning consecutive, simultaneous and efflorescent figures (page 47) produced only chaos that requires an entirely new approach?

In spite of complications I think that gestalt analytical theory is a tight unit and is fully justified and supported by benefits in therapeutic practice: you would expect me to do so since I have bothered to set it all out in this book. Gestalt psychotherapy helps people to become themselves as they know themselves. They grow and they come to live in the zone (page 27). This entails, as Laura Perls wrote –

I am present in what I am doing with no intrusions and distractions, from family, job or other people. I am peaceful, calm, tranquil, positive and only able to make correct decisions. I have no mind other than focus on the single thought. Here I will complete what I am doing. I am only vaguely aware of everything going on around me on that moment. I am courageous and confident[2].

This is where I have been during my creation of writing and as an ending, is me voicing a plea to throw off contaminants and become aware of real gestalt.

Thirty-six thousand dawns
punctuate life lived.
What a song and dance!

Cooperation, democracy and civilization

To create is to remember
Akiro Kurasawa

In a fully fledged democracy all citizens would converse, collaborate and cooperate using voting to come to all group decisions, and to avoid autocratic situations where unelected persons hold themselves superior to others.

At first sight this should be easy to acheive because Tomasello[3] and colleagues have shown that infants of one to three years old naturally cooperate with others and are easily altruistic. Anthropologists have shown that our adult ancestors likewise cooperated and were altruistic in their nurturing and social habits including hunting, gathering and mate finding. The more recent evolution of human social behaviour found domineering people for us, kings, emperors, mafia bosses, C.E.O.s, admirals, and others of criminal propensity who used pseudo-legal systems to legitimize their often murderous activities.

We don't have to tollerate these bastards for ever. We can found a new legal system to equalize all citizens in democratic oportunities. Using Freudian terminology we can say that the people of such new times will be less neurotic and there will be less need for any kind of therapist, including ones thinking and feeling in the gestalt mode.

Democratic freedom of choice is expressed in the therapist's question which can be subject of negociation: 'I have some ideas and suggestions for you. Would you like to hear them and consider acting on them?' 'What do you intend to have for yourself'.

All possible positive attributes characterize civilization. Total democracy – every aspect of civil society initiated by citizens banding together, debating and voting– the Quakers show us how to do it.

A choice of Moral and ethical behavior can be faciltated by meditation on the subject matter of the "Interpersonal Attitudes" box.

Base and Superstructure

We are all stoneagers
living in the fast lane.
Evans[4].

This book, for me, has been constructed like people construct a house with first attention on the foundations. The superficial features, the visible aspects, of the house can then be created.

Most observers have given major attention to what is most readily observable, in psychological terms, the relationship with neighbouring houses, especially effective and ineffective dialogue, and the environment, the field in which the house is built, which has an effect on the architect and thus the design of the house.

Many people have written about the gestalt superstructure, often with no attention at all to the foundations in gestalt figure and ground. The repetitions among such

Interpersonal attitudes

Cooperation	Individualism
I bother. I make care carefully.	I don't care about or for you.
I am interested in you and what you do.	"Fuck you!"
I engage with you in mutual endeavours.	I contact you.
I have codes of ethics and morality.	There are no such things as ethical and moral behaviours.
I enjoy being a citizen in a society.	I am alone against the world. Anyway, "there is no such thing as society" (Prime Minister Thatcher).
I joyfully cooperate in a great network of butchers, bakers and candle stick makers.	I buy whoever I want.
I am prepared to share my accoutrements . with my friends.	I keep what I've got. "What we have we hold!" (Churchill).
I feel expansively and carefully altruistic with everyone.	I have guarded altruism only towards my family.

authors and lack of innovations dismays me. They stray further and further from gestalt foundations with interests in pseudo-philosophical phenomenology and almost any other kind of psychotherapeutic system. It is time for an update, a return to basic observations and ideas.

Fantasy future

As I am placed now, in the country, far from city life, unable to travel, I cannot envisage doing any therapeutic work.

There are many things I would like to do and I list them here to prompt the interest of other people.

- Quite ordinary problems may be amenable to protofigure analysis. "I just can't get on with older men at work."
- The protofigure system may turn out be available for training people of little psychological or psychotherapeutic background, a sort of counselling system.
- In education investigation of protofigures may provide access to the solution to learning problems.
- In social work a protofigure may lead to the solution of a problem. E.g., a mother is not caring properly for her baby. What else is going on?

- In bereavement protofigures may help to lessen the effect of the trauma.
- In memory loss protofigures may help establish connection to supposedly lost memories.
- In depression resort to protofigures may help re-establish grounding.

Where is the passion?

Passion, like art, is always irresponsible,
useless, an end in itself,
regulated by its own impulses
and nothing else
Edmund White,
writing about Mapplethorpe's photographs.

A mundane attitude to figure development leaves the gestalt map as depicted in the Frontispiece. Much is missing. Where is the subject matter of poetry, painting and music. Where are forgiveness, reconciliation[5] and love. Where is passion?

In terms of the categories of the Frontispiece map, stages <1> and <2>, passion is ecstatic interest, an all engrossing appetite, an undeniable impulse and a need so overwhelming as to be not recognized as neediness. Passion is heart beat rate high at stage <3>, in a soft hug at stage <4>, a

On track

I sit here feeling very tired and yet my mind/ brain is driving my fingers to flip over the keyboard.

F Is tiredness a natural feeling or is it an indicator of something else going on?

E [No worry. It's me] I'm just tired, that's all.

F Slow down a bit. Think only of being tired … and then wonder if there is something else going on as well.

E [Yawns] … well, I feel a bit hungry … and thirsty.

F So you feel tired, hungry and thirsty. Is there anything else as you feel tired, hungry and thirsty?

E And peace and quiet … except for you going on and on with spotlessly clean questions.

F So you feel tired, a bit hungry and thirsty while I interrupt your peace and quiet with spotlessly clean questions. What happens next as you feel tired, a bit hungry and thirsty while I interrupt your peace and quiet with spotlessly clean questions?

E Oh piss off, can't you!

F I want you to know that it is easy to type these repetitive phrases using the cut and paste method.

E [Sarcastic voice] How modern for you! I feel stoneagey. Bored, tetchy, tired and I want to go to sleep. So shut up!

F And you think …

whispered "I love you" at stage <5> with tremulous emphasis on "you." Passion is shown by physiology and psychology. Breathing is constrained or eruptive, musculature is taught or flabby, secretions pouring.

On the track of passion again, I feel my energy spring from my underbelly, rising into my hoarse voice, out of my hands and onto this page: originally penned, then word processed and my eyes glisten.

Dispassion?

I am riding the balance point
between juvenilia and senilia.

During the time of creation of this book ideas suddenly appeared for me after quiet periods of gestation while my psychological work with groups, couples, myself and 1:1 continued. The contents of this book have been shuffled as my ideas changed until change, variation as time passed, became the predominant theme.

Construction and deconstruction interwove until, in the last few years, from time to time, the text began to influence me. Writing about depression was creating a metaphor with powerful negative connotations. I had to be alert to my feelings and give myself self-therapy. I often became bored with all this gestalt rigmarole, became gloomy as the prospect of finishing this book seemed to recede into the future, paralysed by the enormity of the project, afraid of bouts of illness of the style that afflicts all old people, including two breast lumps, a rebored prostate gland, diabetes type II and diverticulosis. The doctors discovered recently that I had two mildly effective congenital heart lesions! I also scared myself with fantasies of an unwelcoming reception for this book.

Senility is closing in on me. I have often forgotten the beginning of a paragraph by the time I reached the end.

For the love of a people and of their language I suddenly interrupted work on this book and began to learn Italian. This process illuminated the nature of language for me and I think this book is the better for it.

For the love of my Spirit guide I suddenly interrupted work on this book and wrote another about spirituality and this turned out to be another kind of gestalt essay.

I have never regretted starting on this road. I have always known I could cope, even if I had to take days, weeks, sometimes months away from the project. Nurtured ideas matured in spite of what I might be consciously aware off. Fragments of ideas popped into consciousness while on the beach or in bed, in the hypnogogic state while preparing to sleep. All were instantly committed to note pad or electronic recorder in fear of forgetfulness, and that action encouraged wakefulness rather than sleep. Otherwise, I, of reasonably good memory and not arthritic, am dozing in the spring sunlight in my garden, subconsciously fermenting words and phrases until now, right here and right now.

And did I have an ancestor
"in ancient times"
who was troubled by fear
and by fluttering in his belly
until his worry transformed it
into something else
and something else
until it became courage.

Scientific Gestalt or Gestaltic Science

Progressive and regressive social behaviour

A week ago, as I write, young people in 16 areas of London rioted, looting and burning out shops and being attacked by police. This behaviour was accounted for in many different forms by different people. I looked at it on TV in a remote corner of Dorset. The events showed two forms of regressive behaviour, the rioters in the mode of four-year olds and the police as thugs hired to protect the Lords property, again regressive behaviour as each policeman did as he was told without thinking. The senior police officers behaved as if they were the Lords that they aspire to. It was old fashioned class struggle.

Both sides, rioters and police, despised the vigilantes who tried to protect property and restore order in a truly progressive way. Positive, creative progression is good for everyone involved. Negative regression leaves everyone dissatisfied.

And I am involved too as the "On Track" box illustrates.

> I have coloured my map of the city.
> I have added contours
> place and street names
> parameters and concepts.
>
> I am sure, now
> that my map is of my city.
> I am sure, now
> that I know where I am.
>
> Only you can turn
> monologue into dialogue
> and, after all is said and done,
> that is what this book is all about.
>
> I wonder if you will be so kind
> and let me know
> if this map has any
> relevance for you.

Human kind

Our physiology is probably still slowly evolving though we may not notice it in the context of all that the physicians and surgeons do. Our behavior evolves, again hidden in the context of all our social interactions. We have hope that the poisons of individualism will be diluted out by the passions of social democratic functioning.

Wise we are if that is one of our intentions.

Notes and references

1   Perls, F., 1973.
2   Laura Perls is my heroine – she chose to put up with Fritz Perls! To check her influence on me you can read the seven contributions to gestalt approach she made in 1953, 1956, 1961, 1976, 1992 and 1994.
3   Tomasello, 2009.
4   Evans, 2005.
5   Archbishop Tutu preserved social peace in democratic South Africa and showed all politicians how to do so.
6   Apologies to Blake (1757 → 1627) who was enamored of Jerusalem.

I began Part 1 of this book with quotations from Laura Perls and, for balance at least, I end with another[2] –

> The neurotic is really a person
> who is afraid to cope
> with the process of dying
> and therefor he can't live.
> Being aware of one's own mortality
> is actually an incentive to being alive.

—=(☆)=—

Appendix 1

Clean, Effective Language

The purpose of using clean language
is for the facilitator
to tune in to the experimenter
and for the experimenter
to know that the facilitator has in fact tuned in.

The engagement of the Experimenter and the Facilitator is fostered if the latter communicates using the sub-set of language selected by Grove[1] and termed by him, clean language. This account of clean language hangs, for convenience of exposition, on the skeleton of the protofigure procedure set out in Chapter 10.

Empathy

Clean language is the language of empathy whereby the facilitator does not tell the experimenter what to feel or think or what to do, after the manner of Rogers. The experimenter discovers his intentions while he or she is supported by the facilitator in what is probably a very light trance. The only pressures from the facilitator are in the direction of discovery of a felt-sense (lump in the belly, motor tic, etc.) and then a sequence of gestalt figure stages which are brought to completion and grounding.

Grove's clean language is concerned only with gentle curiosity and suggestion. It is neither aggressive nor coercive. The facilitator using it follows the experimenters expressions and leads forward to new concepts.

Clean language is the facilitatory tool. By using it the facilitator concentrates the experimenters attention, producing a state "of concentrated self-absorption" as the experimenter is interested more and more in his own discoveries. The facilitator does not instruct or intrude in the experimenters process. There is no formal induction of a trance state which occurs anyway, as it does with Erickson's

procedures[2]. More aspects of clean language are –

• The facilitator's first key question is very important in that it must not orientate the experimenter in any particular mode of operation and language, past, auditory, metaphorical, etc. The experimenter sets his or her own scene. Grove[1] lists the following recommended openers for the facilitator; 'What is it that you want?' 'What do you need to have happen?' 'What would be helpful?' 'What would you like?' I find it more productive to ask –

'What do you intend to
have for yourself?'

Or something like:

'You can wonder
what you intend to
have happen?'

• Statements that support the trance state are valuable; 'Take your time to discover as much as you need about that, there, and let me know when you have done that.'

• The facilitator does not need information and does not ask questions that would yield information of interest for the facilitator rather than the experimenter.

• The information created by the experimenter is an agent of autosuggestion[3].

- 'We want the client in *subjective* reality because that is where the change is going to occur. (Grove)
- The facilitator is 'not present in clean language.'
- The purpose of clean language questions is 'to develop the client's internal process.'

All the above indicates that the experimenter is in control of what happens to him- or herself and does not get in the position of saying: "There you go. Pushing me around just like my Dad did."

Development of information using clean, purposeful language

The experimenter is interested in the felt-sense, the protofigure that he or she found and will probably seek more information about it and thus promote empathy. The facilitator, has a purpose, as stated above, to promote the experimentor's activity by talking using the clean language, which was introduced by Grove and Panzer[1] although it resembles a similar subset of language proposed by Levitski and Perls[4]. One may consider clean language to be an esoteric adjunct of therapy. It is, however, the normal, highly effective language of civilised people, though without some of the adjuncts that are of interest during therapy.

Taking up observations from Chapters 2 and 3 and the Frontispiece, where the stages of development of figures are set out. The hypothesis is –

A new gestalt figure
originating from a felt-protofigure
will follow the general course
of figure development,
as summarised in the Frontispiece.

Lumps in the stomach or throat, motor tic in an eye, dark cloud hanging over and many similar things are the felt-senses described by Gendlin[5] and now thought of as protofigures. A rock in the belly can be acknowledged as such and thought to be a fantasy. So it is but Gendlin showed that such an apparition could be changed with diagnostic and treatment advantage. Grove extended Gendlin's observations by being more exacting, using a metaphor analysis system.

The present author, in cooperation with David Grove[6], pointed out that a clearer way of analysing development processes was to use gestalt analysis. The initial observation, a ball in the throat, a forest blocking progress, a funny eye that flickers, was seen to be the precursor of a gestalt figure and was termed a protofigure. Sometimes, for a historical reference, the term felt-protofigure was used.

In the normal course of events a protofigure may be noticed and then forgotten. If it is an eye tic it may bother the person from time to time. Otherwise the protofigure may belongs to a specific occasion – a lump in the stomach justbefore giving a lecture to a thousand people. Or a protofigure may come into awareness when the person quietly meditates on something untowards, like a nightmare or daytime flashbacks from a trauma.

We need to consider the event when two protofigures <1a> and <1b> are found as well as when only one is found, as illustrated in the Track box, page 165.

Beneficial tactics

When carrying put the above procedure, particular attention can, with advantage, be given to the following factors by the facilitator which contribute empathy to the experimenter – facilitator system.

Smooth talking

The first aspect to underline, occurring frequently in the examples in Chapter 9, is the use of "and" or "so" as the facilitatory first word in a sentence. These provide a smooth link to what the experimenter has said. It also helps clarify the situation for the experimenter who may have said something that was not quite what he or she intended to say.

On occasions the facilitator can start a phrase slowly by saying "now", or even "then", as effective grounding words.

Putting questions

The facilitator keeps questions or statements tied to what is going on for the experimenter. Examples are laid out in the clean questions box (page 166). The facilitator keeps them in the simplest verbal form possible, so that, as stated above, a child aspect of the experimenter will be likely to understand what is going on.

Another important tactic is to ask one question at a time. If a question is not answered it could be repeated in a reformulated verbal form or a completely different question is put. An exception to this suggestion is when a double suggestion is employed in the Ericksonian manner[2], to induce a light trance. E.g., 'Would mouse in your chest like to go to your head or stay where it is?'

The feed-back framework

The facilitator follows the experimenters statement that requires feed back by a statement or question of the form –

item for feed back + new question + item for feed back

for example, the experimenter has a toad in his throat – the facilitator says something like – "And you have a toad in your throat. What colour is the toad in your throat?"

This may become complicated as information grows, thus: "And the very small, green, slimey toad in your throat, is it moving at all, the very small, green, slimey toad in your throat."

The alpha state

Grove described clean language as inducing an internal

SINGLE PROTOFIGURE TRACK – use only this track.

Accepting that the protofigure is stage <2> of a gestalt figure (stage <1> was, by definiton, out of awareness) the experimenter's interest can be increased by questions from the facilitator – how large, what colour, etc.

In considering moving on to stage <3> the focus of interest changes from depiction to considerations of energy, activity, – is the protofigure changing? What is happening? Is the protofigure interested in changing?

At stage <4> options and choices become possible. The experimenter becomes more sure of the validity of what is going on in preparation for the consummating conclusion –

At stage <5> the gestalt figure is complete. If the new gestalt is of therapeutic effectiveness this will be known at stage <5>.

DOUBLE PROTOFIGURE TRACK – use both tracks consecutively.

Meanwhile, while still thinking about stage <2a>, the experimenter may become aware of another protofigure at its stage <2b> although this can otherwise happen at stages <3a> or <4a> so there are two protofigures to be dealt with consecutively. The facilitator discusses the situation with the experimenter and invites the experimenter to choose the protofigure that is of principle interest to him or her. Then, having matured that protofigure to stage <5>, as in the first column, the facilitator suggests a return to deal with the other protofigure.

When this second protofigure has matured to its stage <5b> the facilitator asks whether either of the protofigures can move. The experimenter says that one or both can move so the facilitator asks –

What happens when the two protofigures
meet somewhere?

After a short pause the experimenter shudders or makes some other movement and comment on the event, talking about an "electric shock" or something similar at the point of meeting.

The facilitator will encourage the experimenter to enjoy his or her success by expressing satisfaction at stages <6> and <7>. He will also invite the experimenter to assess the events that have just happened by giving attention to the future -

- Observe how you go during the next week and report back, perhaps by telephone. Any symptom, protofigure, etc., will not have recurred.
- Quieten to a meditation state and imagine times in the future when the protofigure might have been expected to occur. I.e., Nights of good uninterrupted sleep are enjoyed. Days pass when a flash back might have been expected and does not recur.

concentration state which he variously described as the alpha state, as not valenced, as not loaded in any particular direction - past, auditory, metaphorical, etc. Questions are put in a way that puts and keeps the experimenter in a 'conversational trance' and is thus not so deep as to induce silence.

A mismatch by the **F** would bring the **E** out of the alpha state and disrupts established empathy.

Reinforcing ideas

Repetitive feed back from the facilitator to the experimenter lets the experimenter know that the facilitator understands and supports him or her in what is going on. The repetition is best provided at the beginning and the end of the proposition so that it sounds natural and unforced. 'And

you have a pain in the back of your neck. How large is the pain in the back of your neck?' Such propositions can be thought about in two sentences though they may not be.

To save space the propositions set out in the box (page 166) are largely set out without the repetitions.

The Facilitator's irrelevancies

The facilitator avoids using 'I need' and 'I want' because they are not relevant to what is going on for the experimenter. Likewise the facilitator does not say 'OK', 'right' or anything else that indicates that he or she is judging, approving or disapproving of the experimenter's presentation. He or she might express thanks as a particularly effective point of emphasis.

Doctors often say something like; "Do this for me." That

**Clean questions for the facilitator to use when seeking
increased information about an experimenter's gestalt protofigure[1, 6].**
In this schedule t"he protofigure seed and developments of it are set as GPF
to represent, lumps in the belly, menacing dark clouds, etc.
N .B. The following questions are set out in abbreviated form. In use
"What colour is GPF" would become something like 'And GPF is there; what colour is GPF
as GPF is there?' See the explanation in the text.,

Characteristics; knowing <2>
And what are you aware of now?
And what colour is GPF (or it)?
And what shape is GPF?
And what is GPF like?
And what is GPF in the belly like?
And does it have a shape or size?
And what is the texture of its surface?
And is there anything else about GPF?
And GPF like what?
And what kind of GPF?
And as you are aware of GPF, where is GPF as
    you are aware of GPF?
And as you are aware of GPF, what is GPF like as
    you are aware of GPF?
And is there anything else about GPF?
And is there anything else about that?

Location; knowing <2>
And where do you have GPF?
And where abouts?
And where could GPF be?
And when you are scared where are you scared?
So when you have GPF, where do you have GPF
    when you have GPF?
And is it on the inside or the outside?

Who? knowing <2>
And who could GPF be?
And how is GPF dressed?

Inviting changes; <2> → <3>
And what would you like to have happen?
And what would GPF like to have happen?
And what happens next?
And what would you like to do?
And what can happen so you don't have to be
scared?
And as that happens, what happens next?
And what happened before that?
And could GPF remember what was intended there
    then?
And what happens next when you don't know?
And what difference do you think this will make?
And how do you know that this will make a
difference?
And what do you think you will be able to do now
    that you couldn't do before?

Inviting activity; <3> → <4>
And as you are aware of GPF, where is GPF as you
    are aware of GPF?
And what would GPF like to do?
And how could GPF do that?
And as you are aware of GPF, where is GPF as you
    are aware of GPF?
And how can GPF do that?
And how can you do that?
And what would be the first thing GPF could do
    so that could happen?
And what does GPF do for you?
And could GPF be interested in becoming more
    energetic?

Prompting and supporting regression
And when … (DG says this question invites
    regression).
And how old could a GPF like that be?
And how is GPF dressed when she is scared?
And how old would you be when you are scared?

Supporting positive affect
And as you grin at the idea of GPF what happens next.

Supporting duration
And how long will that take?
And what happens after it takes that much time?
And how long is forever?
And how long does a hand hit you for?

Supporting inanimate objects
What kind of GPF would that GPF be?
And is there anything else about that GPF?
And what could that GPF be made of?
And what would an GPF like that GPF like to have
    happen?
And how could that GPF help you?
And what could that GPF do?
And would GPF be interested in going to …?

Supporting wonder
And you can wonder[7] how soon GPF will change for you?

Supporting people resources
And who could come?
And what could he or she say to you?
And what would he or she do next?
And where could they take you?

Supporting spiritual resources
And do you have spiritual resources?
And what could spiritual resources do for you?
And how could spiritual resources make a difference?

Choice among options; <3> → <4>
And can / could GPF choose to do something
          different as GPF is choosing?
And as you wonder about GPF can you think of any
way that GPF could happen?

Inviting completion; <4> → <5>
And as you remember that you intended something
do you find that what you intended has
          happened?
And as you remember what you intended to happen
what do you need to do to make that
          happen?
And as you remember what you intended to have
          happen what do you need to do to have what
you intended to have happen, happen?

Elucidatory questions

And how do you know you are (*or* are not) confused?'
'So when you have an ache where is the ache?'
After a quiet pause. 'And what did you discover inside
there?'

Supporting dealing with an assailant
And as he is coming to get you what do you do next?
And what does he do next?
And as someone is coming what does he do next?

Supporting healing
And you have achieved what you intended for yourself.
How will you support yourself as you have achieved
          what you intended for yourself.

Supporting satisfaction; <5> → <6> etc.
And as you remember what you intended for yourself
          do you recognize that you have what you
          intended?
And are you happy with what you intended to have for
          yourself?
And as you have what you intended for yourself are
          you happy that you have what you intended for
          yourself?

Supporting consolidation; <7>
And can you envisage being in a situation out in the
          world where you will know that you have
          achieved what you intended for yourself, and
          find that you have indeed achieved what you
          intended? (That sounds better when the
          achievement is named rather than
          generalized).

is not on. The experimenter is encouraged to be responsible for self. Likewise never ask "What's on you mind?" or "What do you think?" because anything to do with "mind" and thought may not be the subject of the experimenters' attention. Likewise, "What do you feel? because the experimenter may be thinking. "You know" has the listener wondering what he or she is supposed to know and politely pauses, waiting to be told. "I mean …" indicates that what the facilitator said recently was unclear and he needs to be both clear and definite.

Maintaining engagement

In addition to the above points the following verbal forms help to maintain and improve the facilitators engagement with the experimenter –

- Avoid negativity. Be positive when formulating suggestions and questions.
- No aggression, imperialism and domination.
- No passivity; the experimenter needs a friend.
- Be gentle and polite.
- Use the verb form and tense that a child would use .
- Use 'you' when the client says 'I'.
- Use 'he' or 'she' when the client does.
- Use gerund forms of verbs to emphasize a continuous flow and development of activities.

- Use 'happen' to move the process on.
- 'What's happening now/next?'
- 'What happened before that?'
- 'What happened next or before that?'
          (Hypnotic form, Erickson[2]).
- The experimenter's language forms often change during a therapy session to become more like the facilitator's clean language with the advantage of increased clarity. The facilitator does not encourage this change.

Progress: moving on

The experimenter may tend to stay at one stage of the process rather then proceed – the experimenter's concern with the black cat, page 145, is an example. It is then up to the facilitator to move the process on, to move along the figure stages to develop the gestalt. These suggestions may not be applied dogmatically – the experimenter has produced gestalt figures all his or her life and will instinctively know how to do this new one, derived from the protofigure.

While developing the gestalt figure stages there are two aspects for attention –

- Having reached a stage, come to know as much about it as possible
- What needs to be done to move on to the next stage?

Protofigures in the gestalt figure context

A felt-phenomenon is recognized;
a sensation of a lump inside
or cloud outside, etc.,
designated **X**. This is   found as   **stage <1>**
soon as description is attempted,
perhaps in answer to a question;

'What's **X** like?'
Attention moves on as **F** forms more questions to build up specifications for **X**;   **stage <2>**
'What shape is it?' flat, bulky, geometry, spherical. 'What colour is it?'
Attention is given to surface texture, solidity or hollowness, type of material – rock,
 metal, wood, etc. Concrete or jelly, a liquid, emulsion.

The question 'Would **X** be interested in changing?'
helps attention to move on to form   **stage <3>**
where movement may be noticed, with a degree of energy;
'What would **X** like to do?' 'Does **X** say anything?' 'Is **X** interested in moving?'

Then come questions concerning options;
'Is **X** what you intend to have or is there something else?'   **stage <4>**
'Do you want to choose among possibilities here?'

Moving on, 'does **X** feel complete now?'   **stage <5>**
'Have you got what you intended to have?'

'How do you know that you have what you intended to have?'
'Can you imagine a time in the future when you might have been troubled by that trauma or
problem and you realise that you are not troubled by it?'

Expressions of satisfaction
                                                                       stages <6> & <7>

Development of meaning

> **Beware!** A feeling in the
> belly may not be a protofigure
> but be an ulcer.

The facilitator leads the experimenter to discover more and more about the experimenter's discovery, dog, frog, log, by continually seeking more information, prompted by clean questions. The examples in Chapter 9 illustrate the process.

In addition to information for stage <2> and the other stages, excitement and increased energy are sought for stage <3>, options are sought for stage <4> and completion sought for stage <5>. Specimen clean language phrases are set out in the clean language box (page 166).

Further illustrations of the protofigure development processes are set out in the protofigure box above where emphasis is on comparing the relationship of the observed behaviour of the experimenter with the required character of the gestalt figure stages.

Certain words and phrases are particularly valuable for the facilitator in moving the matter on –

Like: 'What is the ball in your stomach like?'

Wonder: 'Now that has happened you can wonder what might happen next, now that has happened.' This exploits the trance like tendency when wonder is used – an awesome way of moving things on.

Can: a gentle way of suggesting movement. 'You can see if it moves.'

When: And when that happens , what happens next.' A gentle way of inviting time change.

Time: 'So take some time to discover what happened before that' slows the experimenter's processes.

Anthropomorphization: this is pretending that inanimate objects had interests and intents – Grove found this to be a valuable extension of metaphor usage. 'And would hollow ball like to move somewhere?' 'And does the rock in your throat intend to do something for you or say something to you?' (There are other examples in the single track description, page 165, and in the proto-figure box on page 168).

Age: It is sometimes useful for the experimenter to have attention drawn to the age of a memory. 'And how old was she when the old man came?'

Each of the facilitator's statements or questions is related to the experimenter's previous answer and, in a mirroring way, repeats and reinforces aspects of the experimenter's answer and also suggests further leads.

Recognizing completion of the gestalt

The mood of the experimenter changes when completion of the gestalt figure is achieved at stage <5> – remember, as remarked above, that the experimenter has, probably without knowing it, been completing gestalts all his life. The experimenter may become obviously happier, may talk about finishing the session and/or may express satisfaction at what he has achieved; stages <6> and <7>. The facilitator supports this eventuality and may encourage it.

The facilitator will know that not only is this gestalt complete but something to do with the initial trauma or problem has occurred – for a possible explanation see the neurology box (page 154).

Future projection

The last phase of protogestalt therapy involves checking that the initial trauma or problem has indeed been solved. The ways of doing this are set out at the end of the Track box.

It is sometimes not possible to deal with all aspects of a trauma or problem in one session. The process of finding protofigures and developing them can be repeated until the experimenter is completely satisfied.

By using *clean language* the facilitator facilitates the experimenters process of discovering a new gestalt. The facilitator helps the experimenter to organize and format this information in a useable, intelligible form. Both experimenter and facilitator gain and interpret knowledge about the growing gestalt.

Notes and References

1   Grove and Panzer, 1989.
2   Erickson, 1976.
3   Couè, 1992.
4   Levitski & Perls, 1970.
5   Gendlin, 1996.
6   Grove, personal communication.
7   Wonder as an amazing word, see page 173.

A rather shy proton
roamed out one night
encountered electrons
became ruddy bright.

—=(☆)=—

Appendix 2

Practical Matters

> As on earth so in heaven
> As in heaven so on earth
> A very, very old saying.

The Signs and Symptoms of Health[1]

These points are referred to on page 6.

HEALTH is perfection my of body organisation.

HEALTH is my intellectual energy.

HEALTH is the fullest expression of all my faculties and passions acting together in perfect harmony.

HEALTH is freedom from pain in my body and freedom from discordance in my mind.

HEALTH is my beauty, energy, purity and happiness.

HEALTH is that condition on which I am the highest expression of the power and beneficial activity of nature.

When I am perfect in my own nature, body, mind and soul, each perfect in their harmonious relationship with the others, and I am perfect in my harmony with all nature, including with my fellow women and men, I may be said to be in a perfect state of HEALTH.

ROUTINE PROCEDURES

All routine procedures for problem solving are figures and some will be set out here to have stages that correspond with the generalized stages set out in the Frontispiece. A stage or two may sometimes not actually be in evidence, as stressed earlier, and this would probably be due to speed of execution of such a stage. Some figures may involve repetitions and can be represented as if with internal short circuits.

The procedures set out in first person language can be recorded on tape by you and played back to guide you. Otherwise they can be read for the benefit of another person.

The relaxation, meditation and other quiet procedures may cause you to fall asleep. You evidently needed to do so. When you awaken you can start the procedure again. Where there is reference to a breathing rate it is a suggestion that you set to about 8 complete cycles per minute as recommended by Cluff[2] and on page 101.

Centring and awareness

An energizing experience

This procedure derives from Goldstein.

I use my Hara as an energy source. I am right handed so that my dominant hand is my right hand and my tonic hand is my left hand. The opposite pertains for left handed people.

This procedure works well, i) for energizing and centring and ii) as a guided fantasy in which only the movements of breathing occur. While this description relates to standing you can, with practice, do it while lying down.

1   I stand with my arms hanging loosely, hands by my sides.

2   I stay aware of my breathing and make all subsequent movements in time with my breathing; outward, expansive movements in time with my in-breaths and inward contractile movements in time with my out-breaths.

3   I put my dominant hand over my heart and become aware of my heart beat and of my breath flowing into my chest. I put my tonic hand[3] over my lower belly (hara) and become aware that my breath also flows into my belly. I set my breathing rate at four heart beats for my in-breath and four heart beats for my out-breath.

4   I count my complete, in plus out, breaths.

5   After I have completed the last section four times I end on an out-breath and I hold my breath to a count of four beats while I move my dominant hand on to my hara and my tonic hand to rest on the centre of my chest.

6   On an in-breath, to a count of four beats, I open my arms widely with my hands at shoulder level. On an out-breath, to a count of four beats, I reverse and return my hands to my hara and my chest, while maintaining my erect posture.

7   After I have completed the last section four times I end on an out breath and I hold my breath to a count of four beats while I move my hands so that my dominant hand is on my hara and my tonic hand is on my stomach.

8   On an in-breath, to a count of four beats, I open my arms widely with my hands at stomach level. On an out-breath, to a count of four beats, I reverse and return my hands to my hara and my stomach while maintaining my erect posture.

9   After I have completed the last section four times I end on an out-breath and I hold my breath to a count of four beats while I move my hands so that my dominant hand is on my hara and my tonic hand is on my neck.

10   On an in-breath, to a count of four beats, I open my arms widely with my hands at the level of my neck. On an out-breath, to a count of four beats, I reverse and return my hands to my hara and my neck, while maintaining my erect posture.

11   After I have completed the last section four times I end on an out-breath and I hold my breath to a count of four beats while I move my hands so that my dominant hand is on my hara and my tonic hand is over my eyes.

12   On an in-breath, to a count of four beats, I open my arms widely with my hands at the level of the top of my eyes. On an out-breath, to a count of four beats, I reverse and return my hands to my hara and my eyes, while maintaining my erect posture.

13   After I have completed the last section four times I end on an out-breath and I hold my breath to a count of four beats while I move my hands so that my dominant hand is on my hara and my tonic hind is on the top of my head.

14   On an in-breath, to a count of four beats, I open my arms widely, directly above my head and with my eyes open, looking straight ahead. On an out-breath, to a count of four beats, I reverse and return my hands to my hara and my head, while maintaining my erect posture.

15   On an in-breath, to a count of four beats, I slowly bring my hands down to rest by my sides where they began. I notice my exhilaration and my sense of satisfaction and I continue with something completely different.

Stopping recycling

This procedure is referred to on page 136 .

If, when preparing to sleep at night, or on other occasions, ideas flood into my mind in a sequence that recycles, on and on in the hypnogogic state, I say:

STOP!

I deliberately choose to think about something else.

Focused awareness: the paradise place

This procedure is referred to on page 68 note 5.

<1>   Become as comfortable as you can, shut your eyes and become aware of your inner sensations, your breathing and your heart beat.

<2>   Let your mind wander and become in contact with the most beautiful place you have ever been to. Stay there as long as you please, developing your awareness of what you can see, hear, smell and feel.

<3>   Keeping your eyes closed, switch your attention to this room and everyone in it. Look around carefully, fill in more and more detail. See, hear, smell and feel.

<2>   Switch your attention back to the most beautiful place again. What can you see, hear, smell and feel? What has changed for you compared to your experience in stage <1>.

<3>   Open your eyes and look round this room. See, hear, smell and feel. What has changed for you compared

to your experience in the previous stage <3>.

<5>   Share your observations with your fellow group members.

Sensitizing awareness of the figure stages

This procedure is referred to on page 47.

When teaching the value of awareness of the stages of figure development I found I needed experiments for students to conduct. The following procedures were successful.

T hides a lemon in his pocket. He asks the students to become comfortable, relaxed and each aware of his or her breathing. He then says something like –

T   I am going to produce an object for you to see. You can, if you like, be aware of what happens for you in minute detail.

T then produced the lemon and said: "Here!"

After two or three minutes **T** invites the students to share their experience in pairs, followed by full class discussion.

The outcome was that all the students reported immediate mouth sensations and salivation, stage <1>, followed rapidly by knowledge that these were happening, stage <2>. Then came a degree of energization, usually fidgeting in the chair, and various impressions including revulsion: "I don't like to suck lemons"; increased interest: "I like to use lemons in cooking" and "lemons go well in tea."

These were all stage <3> degrees of excitement and affect. Nobody was interested in holding or doing anything with the lemon.

A red apple was not as evocative as a lemon. A small tray of chocolates produced the activity of eating followed by satisfaction.

The other experiment the students were encouraged to experience was the use the familiar phrase: "Now I am aware …" in contexts where it would be easy to sort out the figure stages; the meal table being an obvious place.

Solving problems

Metaphors for metaphors

Accepting that metaphorical stories are effective curative agents as Erickson showed and more recently presented by Yapko I wonder about metaphors in everyday life. If a person tells him or herself stories about his or her life style they can be destructive if negative in content and the curative move is to reformulate such stories to make them positively effective. The person can then realize the benefit of setting out to create metaphors for self that are always assertive and with strong positive endings.

A metaphor for this kind of positivity would be something like: My Grand daughter, Kimi, aged 5, wanted a bed time story so I wondered about some of the Amerindian legends. Coyote the wise trickster seemed to me to be the ideal central character. He was always creating something marvelous. He found the Medicine Wheel out on the prairie. He examined his world carefully before becoming busy. He was friendly with everybody around him, Bear, Wolf, Rat and even People. Kimi, aged 12, wanted to become a Doctor – a real physician.

The dog in the yard

A good guard dog gives major aggressive attention to people who arrive and are not known to the dog. If the boss man comes out and introduces the dog to the new arrival the dog learns to accept the new arrival as one of the family.

Memory boosters

This procedure is referred to on page 33.

Positivity: my memory is part of me – I trust it, I treat it gently and politely. I thank it for its gifts. To emphasize; whenever my memory gives me what I want I say "thank

you memory:" This is at the end of a memory figure at stage <7>.

A procedure to use when memory seems to be asleep is the self-hypnosis concept involving the word *wonder*.[4]

I wonder how soon I will remember this person's name, in twenty seconds or a minute. The name appears for me within my next sentence or so, as I go on talking. Another person may have a visual or auditory impression. This procedure can be used for all sorts of memory problems; "where did I leave my watch?" "where was I going when I left the kitchen?" being popular examples. Then say, as emphasized above, "Thanks memory" – even if you keep the words silently to yourself.

This memory boost procedure is a simple conjunction of "I wonder" and a time interval. The latter is essential for gently asking for information and not waiting years. I have had group members translate this invocation into French and German with beneficial effect.

Wondering is a very useful process and you can wonder if, when you wonder, you are questioning or exclaiming or both? There are many examples of effective wondering in this book.

"I intend to do so-and-so and be complete in five minutes" has a similar powerful effect. Again some form of intent is expressed together with a time limit.

Become Calm

This procedure helps when dealing with mild traumas.
- Become as calm and quiescent as you like to be.
- Imagine that the original trauma was photographed while it was happening, producing "snap shots".
- Now look at these one by one.
- If any trauma reaction appears retreat to the

- excitement that preceded the reaction and then to the calm state that you started in.
- Think about what you would be doing right now if the event in the snap shot was happening now.
- You remain quiet and without emotion.
- Carry on looking at the snap shots until you produce no emotional reactions at all.
- Look again at some of the earlier snap shots that had produced a reaction and enjoy your calmness

A "Wolpe" method, arachnophobia

Wadd (W) complained of arachnophobia. He knew its name and had read a lot about it and still dreaded meeting a spider on the stairs. After relaxation, meditation and focussing on a fantasy trip (page 175), **F** suggested that **W** imagines seeing a minute money spider hanging on a long thread and then return in imagination to his favourite place. This did not produce an emotional reaction. Each challenge was subject to the same intellectual evaluation, how scared is it of you, a large human? Will it attack you? Will it penetrate you? Is it particularly dangerous because it may bite or carry dirt and contamination?

The next spider encountered was two or three mm in size with the same lack of reaction and return to meditating. And so on by steps of size up to spiders that were 20 to 25 mm across and very hairy. This produced a small emotional reaction that dissipated after 2 or 3 minutes. Each increase in spider size thereafter produced a small reaction which dissolved with time. After a final 5 minutes in meditation without spiders the patient was aroused to full awareness and appreciation of the room he was in and then asked to imagine that there was a spider on his knee. He had no reaction and pronounced that he now liked spiders. A few days later a meeting with a spider in his garden interested him but he did not try to touch it. There is another point here: touching a spider puts a human smell on it and that may antagonize another spider.

In gestalt terms each approach to stage <5> was made with diminishing, controlled affect thus depriving the psychovirus of its energy.

A CCTV method, panic and pyrophobia

This account is a disguised version of what happened to **Xeno (X)** whose trauma began one evening on the way home. He walked from work to Kings Cross station on the London Underground rail system and as he descended the escalator noticed a strong smell of burning. He took little notice of it because the underground passages often had very unpleasant smells.

When he got home he watched the TV news and saw scenes he had just traveled through and heard that a disastrous fire on the escalator at Kings Cross had killed a great many people. He felt his "blood run cold" as he told his wife and children of his near miss calamity.

Subsequent nights he traveled home by a route avoiding Kings Cross and it was not until some months later that that station was reopened and he used the familiar escalator again. He had a strong panic attack with sweating, heavy breathing and he began to run down, as well as he could, through the crowd of people. At the bottom he ran to the train platform and felt so dizzy that he sat on the floor, back to the wall. After several trains had passed through he got on one and continued to go home although he was still trembling and had difficulty walking and standing in the carriage. (Nobody took any notice of him and he reported speaking to no one). Other journeys on other days in the underground produced panic attacks when they involved use of an escalator.

In the therapy room he emphasized the nearness of his involvement in the fire – a few minutes he said but it was actually probably 15 or 20 minutes. He also stressed his feelings of horror when he saw TV and newspaper pictures of the fire.

The sequence of depotentiation of his trauma was as follows –

- He was encouraged to follow instructions and to nod his head gently each time he reached completion.
- His attention was drawn to his favourite place and he was invited to imagine that he was there, seeing the place, smelling the smells of the place and knowing how he enjoyed being there. He took about 10 minutes before nodding his head. (This processes establishes a strong observing figure and a positive protofigure that is available at any time).
- He was encouraged to find in his favourite place a complete closed circuit television (CCTV) equipment, camera, batteries, video tape player, monitor screen and remote control pad and all set up so that he could see himself on the screen, looking at the screen. (The process of dissociation).
- He was invited to play a tape of the fire, to watch himself watching the screen and to switch back to enjoying his favourite place if he needed to. He used the control pad to run the video forwards and backwards. (Running the deleterious event protofigure and figure).
- At first he showed small signs of trauma, sweat on his forehead and trembling hands. He switched his interest between the fire and his favourite place several times.
- When he was no longer showing any signs of trauma he was invited to imagine a time in the future when he would go down the escalator at Kings Cross and describe what was happening for him.
- He described going down slowly, standing on the right, remembering that there had been a fire and that, at that time in the future, there was no chance of a fire because engineers had repaired all the equipment. He did not mention any anxiety feelings and certainly no panic. (The trauma memories of the limbic system had been replaced by rationalized memories from the neocortex. The deleterious protofigure had been deleted from memory; page 154).

- He was invited to spend a few more minutes visualizing and "living in" his favourite place before returning to full realization of himself in the room with the therapist and other people present.

X reported by telephone that he was now traveling relatively comfortably in the underground and was not feeling untoward anxiety.

Anyone interested can realize this sort of thing for him- or herself with suitable care and attention to detail.

Anti-anxiety

Stand out of the Way
The Tao

This procedure is referred to on page 100 and is for instant amelioration of anxiety for anyone who can remember to use it on the right moment. It is no substitute for longer psychotherapy.

<1> **STOP**:
become aware of your present feelings which you call anxiety or psychic pain or depression, etc.
Remember the excitement that preceded these feelings

<2> Yawn and stretch

<3> Breathe out as far as you can and then tighten your chest wall and belly muscles a little more. Hold this tension. Breath in and continue to be aware of your breathing rate.

<4> Be aware of everything that happens as you hold your breath – ideas in mind, visual and auditory memories, fantasies, hallucinations. Make positive decisions about them, e.g., if the fantasy was of your father shouting at you and hitting you, decide that he can only be gentle with you now.

<3> Again, remember the excitement that preceded your anxiety. Get in touch with that excitement. Feel it once again.

<4> Do something next; feel your personal freedom as you add joy to your excitement.
If, instead of joy, you feel fear – stay with the fear and get to know what it is all about and then go back

<3> to your excitement stage.

<4> If, instead of joy, you feel anger – stay with the anger and get to know what it is all about and then go back to your excitement stage.

<3> If, instead of joy, you feel mixed feelings, jealousy, guilt, etc., – stay with them and get to know what it is all about and then go back to your excitement stage. Go back to <2> and re-cycle until you reach here again

<5> and are happy.

<6> Express your satisfaction.

<8> Become ready for whatever happens next.

Relaxation, meditation and fantasy trips

The most relaxing, comfortable, productive, situation occurs for me when I have a single focus of attention and am then concerned with one simple, quiet figure only. The following procedures, in my experience, achieves that happy state of gentle, assertive concentration You can, if you choose, flow in without drive or effort.

Relaxation[5]

This Jacobson procedure is referred to on page 90.
There are two ways of starting this process. If **X** reads this slowly to himself or has recorded it for himself he starts at § below. Otherwise if **X** reads it slowly to **Y**, **X** starts on the next line and **Y** repeats the words silently or aloud.

You can repeat after me and to yourself what I say and my I becomes your I and my my becomes something important to you. I am interested in my processes. I am I.

§ I attend to my breathing, counting to myself if necessary to ensure that the in-breath takes as long as the out-breath.
And all the movements I can make in the following activity can be in time with the movements of my breathing.

I am aware of my feet. I tighten the muscles of my feet, hold the tightness and let go, becoming aware of the soft calmness of my feet.

I am aware of my lower legs, I tighten the muscles of my lower legs, hold the tightness and let go, becoming aware of the soft calmness in my lower legs and of other feelings which I will remember for my future attention.

I am aware of my hands, I tighten the muscles of my hands, hold the tightness and let go, becoming aware of the soft calmness in my hands.

I am aware of my lower arms, I tighten the muscles of my lower arms, hold the tightness and let go, becoming aware of the soft calmness in my lower arms and of other feelings which I will remember for my future attention.

I am aware of my upper arms, I tighten the muscles of my upper arms, hold the tightness and let go, becoming aware of the soft calmness in my upper arms and of other feelings which I will remember for my future attention.

I am aware of my belly, I tighten the muscles over my belly hold the tightness and let go, becoming aware of the soft calmness over my belly and of other feelings which I will remember for my future attention.

I am aware of my lower back, I tighten the muscles of my lower back, hold the tightness and let go, becoming aware of the soft calmness in my lower back and of other feelings which I will remember for my future attention.

I am aware of my chest, I tighten the muscles of my chest hold the tightness and let go, becoming aware of the soft calmness in my chest and of other feelings which I will remember for my future attention.

I am aware of my upper back, I tighten the muscles of my upper back, hold the tightness and let go, becoming aware of the soft calmness in my upper back, and of other feelings which I will remember for my future attention.

I am aware of my shoulders, I tighten the muscles of my shoulders, hold the tightness and let go, becoming aware of the soft calmness in my shoulders, and of other feelings which I will remember for my future attention.

I am aware of my neck, I tighten the muscles of my neck, hold the tightness and let go, becoming aware of the soft calmness in my neck, and of other feelings which I will remember for my future attention.

I am aware of my face, I tighten the muscles of my face, hold the tightness and let go, becoming aware of the soft calmness in my face, and of other feelings which I will remember for my future attention.

I am aware of my whole body, now relaxed and completely out of awareness yet whole and solid and me.

The heart beat relaxation[6]

This procedure is referred to on page 167.

I sit or lie in a meditation state

I use a hand to find my heart beat in its proper place under the lower margin of my left rib cage. I check my breathing rate and depth.

I move the sensation of my heart beat from its proper place to some other place. When learning I use the following sequence, taking as much time as necessary in a relaxed manner, to move the impression of my heart beat –

From my left chest area to my right chest.

From my right chest to my lower stomach

Back to my left chest

Up to my throat

On to my right shoulder, then to my right elbow, then to my right hand

Up to my throat

On to my left shoulder, then to my left elbow, then to my left hand. Then on both my hands.

By a stepped sequence in each of my legs and then both my feet.

Back to my left chest

On my hara.

On various regions inside my head

After several practice sessions it becomes easy for you to move your heart beat sensation from its home in your left chest area to any part of your body, outside and in, in one jump. This is very useful when relieving pain: move the "heart beat" to that area: head aches, persistent itches, but not pain due to an ulcerous tooth – a dentist relieves that. Always, as the last move, move your heart beat sensation back to your left chest area.

The Observatory[7]

You have a favourite way of relaxing into relaxation and you can do that right now; become calm, tranquil, quite quiet, and peaceful. And it may be that you will help yourself by remembering other times when you have become perfectly relaxed.

In remembering those other times, it could be that you use your minds eye to see the place where you were so happy and relaxed. It could be that you use your minds ear to hear the sounds around where you were so remotely relaxed. Perhaps you can use your feelings memories to be in touch with all the feelings of that time of supreme relaxation.

And while you are considering the ease of relaxing into relaxation you can also begin to consider the use we will put the state of relaxation to and then you find yourself in the open air by a heavy farm gate. You open the gate, pass through and close the gate behind you.

I wonder if it can be a field in early summer time that you remember or if you will create such a field in your minds imagination. Let the grass be lusciously green and let the early summer flowers crowd in bunches across the expanse of the field. Large daisies over here and buttercups over there. Dandelions by the stream and willows hanging over it.

Wander around the field for a while. Enjoy what you see. Make the colours of the flowers brighter in your minds eye so that you see a sea of richness. Notice that birds are singing or twittering, you can hear them with your minds ear. Make the scent of the flowers even stronger.

Now as you wander across the field you can notice a gate. It is the usual country farm gate, wide enough to take a tractor and high enough to keep the cattle in or out.

You can approach the gate and notice the simple pattern of wooden boards that construct it, four across and one diagonally down from the hinge, five bars in all.

I wonder if you will climb over the gate or swing it open, pass through, and close it behind you.

In the next field you can notice that the grass is quite different. In fact this is not a field but open hillside and a little way away sheep are grazing. That is why the grass is cropped short and the tall flowers are absent. In the grass are short stemmed flowers like daisies and vetch.

A path leads from the gate across the field and you can gently follow it, curious to see where it goes, curious to see what is over the brow of the hill.

Soon, as you walk across the turf, you see a hedge. It is tall and consists of hawthorn and hazel. In the hedge is a gate and the path you are on goes right to the gate. So you reach a narrow wicket gate; hear it squeak as you pass through.

And there inside is the most beautiful garden surrounding

an old house. So you wander around the garden enjoying all the flowers you like most, passing a little pond with gold fish and while you are enjoying all this I will be silent until you nod your head.

[pause]

You can then decide to take a brick paved path towards the house, admiring the sight of the flowers as you go, catching the various scents of the flowers and hearing the small birds sing.

There are three steps up from the garden path to the porch and the front door. You can ascend the steps, curious to know what the inside of the house is like.

The front door is open so you know that you are welcome. In fact the mat inside the front door has the proverbial statement on it – "welcome". So you can go from the warmth of the garden into the cool of the house. It is dim inside and it can take a little time for your eyes to adjust to the level of light that filters in through high windows.

While you pause to adjust to the light you can become aware of the smell of the house and you can wonder if you are noticing the fragrance of flowers in vases or are noticing the smell of polish on the wood work of the stairs or even something else.

As you look around you can notice that all the doors have neat labels on them. There are stairs going up and stairs going down. And you can know that if you go down stairs you will find rooms marked fear, anger, bother, sadness and such.

On the ground floor, where you entered, are rooms marked food, drink, TV, Carpentry, metal work, pottery and flower arrangements.

Up the stairs are rooms marked peaceful rest, music, art, creation and your hearts desire.

You can sense that there is a rule of the house and to follow the rule you can go down the dark stairs into the basement.

The rooms of this lowest floor are marked fear, anger, bother, sadness and such. Chose a room to visit.

In the room marked "fear" is a great furnace. As you open the door the furnace door opens and you see a great maw of orange fire. The fire consumes your fear and warms your gut – you can feel the energy flowing in your belly and flooding round you body, filling your legs, your arms and your head with the energy of fearless life.

In the room marked "anger" is a large stone sink with a tap that continuously runs clean, clear, cool water. There is also strong carbolic soap and rough scrubbing brushes. You can scrub away your anger and flush it down the sink.

In the room marked "bother" you can find a thick rope, fastened somewhere away in the future. Take hold of the rope and walk, pulling it from hand to hand. The rope leads you into another room and then into another and finally into the room marked anger. Scrub your self with carbolic soap and leave.

In the room marked "sadness" is a pool of water set amid flags stones. The pool looks black, deep and bottomless. Sit at the edge of the pool, lean over to see the bottom. See your reflection in the pool. See your reflection shattered by rings of disturbance as your tears drip into the water. Stay until dry.

While you explore the rooms of the basement I will be silent until you nod your head.

[pause]

Well … now it is time to ascend the stairs, out of the basement into the clearer light of the main hall way. As you reach the hall you can know that the rooms there are marked food, drink, music, carpentry, metal work, pottery, TV and flower arrangements. You can even find special rooms dedicated to your special needs.

Chose a room to visit. Inside you will find everything you need to create, whatever you would like to create, be it food, wooden things, pottery, flower arrangements or what ever.

While you are busy creating creatively I will be silent until you nod your head.

[pause]

Well … now it is time to put away the tools you have been using and you can leave the room in which you have been working.

It is time to go up the stairs and you can know that the rooms in the corridor at the top of the stairs are marked peaceful rest, music, art, creation and your hearts desire.

You can chose one of these rooms to visit and I don't need to tell you what you will find there.

While you experience peaceful rest, music, art, creation or your hearts desire I will be silent until you nod your head.

[pause]

Well … it is time to be doing something completely different so you can finish what you were doing on the first floor of the house. You can leave that room and return to the main stairway where your attention is taken, on looking down, by the glare of sun on the garden as seen through the door-way.

But the stairs still go up, so you ascend again and eventually, at the very top, you find an observatory, a small square room with windows looking out in the four directions, north, east, south and west.

You can look out over the garden seeing the path that led you in, look out over the fields, the hills and far away take a glimpse of the glistening sea. As you gaze entranced you can know that you are looking at an overview of your life, all you have created and all you can achieve for yourself in the future. As you meditate on these revelations I will be silent until you nod your head.

[pause]

Well, … it is time now to descend into the body of the house and you can discover that you had ascended by the back stairs. You can now go down the grand staircase, gracefully, fully enjoying the feeling of ownership of the house.

In the majestic hallway you step over the mat marked "welcome", go down the three steps and out into the garden. While you wander, taking pathways that are different from those you used earlier, you can admire all that is beautiful for you in a garden, passing a luxuriant lake and I will be silent until you nod your head.

[pause]

Well … it is time now to realise that there is a great wide

world outside the garden and you become curious to know where the sheep have wandered to, on the hillside.

So you leave the garden through a wooden gate that is different from the one by which you entered and it does not squeak as you pass.

In the field you pause and see that there are now cows instead of sheep – oh! they have such a relaxed life!

So you follow the path back across the short cropped field and you can wonder if it would be possible to count all the daisies that sprout in the grass.

At a five bared gate I wonder if you will climb over or open it, pass through and close it after you again.

In a field of long and luscious grass you walk along the banks of a stream that sparkles in the sun and burbles gently as it passes. There, at a heavy farm gate, you can know that this is where you started from and it is a very suitable place to leave.

You can thank all the parts of you for co-operating together in your present endeavours. Ask all the parts of you to cooperate together in your future endeavours.

I wonder how much you remember of your adventure and I wonder how much of what you remember you would like to share with your friends, here and now.

Reactivate slowly in your own time, becoming aware of the seat under you and of the floor under your feet. And as you open your eyes become aware of this room, of yourself in this room, of me here talking to you and of all your friends also becoming aware of this room and of you sitting there.

The obstructive power of negativity

Reference was made on pages 28, 97 and 145 to the deleterious effects of negativity, particularly in the context of suggestions and commands. The following essay is in justification of this stance.

At first sight such negatives are simple deflective interruptions to contact, are avoidance processes of the form described by Polster and Polster[9] although these authors did not discuss negation as such. The examples and results of experiments to be discussed here are mostly from experience outside a therapy situation and the implications for gestalt therapy will be discussed. The interim hypothesis is that interpersonal contact is greatly improved when the use of negative commands is abandoned.

Observations

1. A three year old child habitually threw its food around at meal times. "Don't do that" and "Don't throw your food" were ineffective as instructions from mother or father. "Keep your food on your plate or eat it" was effective and peace was restored.

2. I was waiting in a short queue in my bank one day behind a woman who was talking across the counter to the clerk. Between us was a push chair carrying a child, probably a boy of about 2 years. By the side of the push chair stood another child, probably a boy, of about 4 years. I could see clearly that the 4 year old was being very gentle with the 2 year old, patting his hair and cooing; they seemed very happy. Mother could evidently feel the push chair vibrating and occasionally glanced over her shoulder at the boys. Suddenly she shouted to the 4 year old: "Don't hit him!" The 4 year old promptly hit the 2 year old, jumped back, and then ran giggling hysterically to the back of the bank.

3. A young teacher told me about an incident that happened with a group of school children in a swimming bath. The children were leaving the water to return to the locker room and ran dangerously on the slippery edge of the bath. One teacher called out "Don't run" and the children continued to run. Another teacher called out "Walk slowly" and all the children responded immediately by walking.

4. A nurse told me about a woman with a small child, who was probably about 5 or 6 years old, whom she saw standing on the curb separating the foot path from a road busy with traffic. The woman said to the child: "Don't run in the road!" The child ran straight out, began to wave its hands in the air and ran back unharmed, seeming to be very confused.

5. Another nurse told me that she worked with senior citizens. One of the jobs she did was to take the infirm ones, one by one, to the bath. She noticed that if, when she wanted to cross the room to get a towel, she said: "Don't get out of the bath, I'm going for a towel", by the time she came back the patient was climbing precariously out of the bath. If she said "stay where you are, I'm going for a towel" the patient was still quietly semi-immersed when she returned.

6. A client reported in group that he had difficulty in getting up in the morning. "What do you say to yourself as you waken?" "I mustn't stay here. I can't stay here any longer, things like that". "Are those messages effective for you?" "No". "What could you say that could be effective for you". After some hesitation: "I could tell myself to get up." "Could you experiment with that message to yourself?" Some weeks later he again raised the subject of getting

up. "I'm much better at starting my day, now. I tell myself what I want to do and get on and do it, which involves getting out of my bed".

7. After some feed back in the group (example 6) about getting up in the morning another man said, while laughing: "I was in the tube last week when a wasp came by. It got around the hand of the lady next to me and I said: 'Don't move'. She moved her hand and accidentally caught the wasp between her fingers and it stung her".

8. This is to exemplify many occurrences in group of rather similar internal messages about memory. A young man said: "I was going to bring a poem I wrote on Sunday but I forgot. I don't remember things these days." "Is that your message to your memory, 'I don't remember?'" "Yes, of course. That's how it is". "Is that the way to encourage your memory, tell it it doesn't remember?" "That's just how it is". "Could you think of another, more effective way of dealing with your memory?" "Well, you told me before, I could be pleased every time it remembers and tell it that it does remember. I have to be able to remember to do that".

Experiments

The following experiments were set up in classes of students studying counselling skills. There were approximately 80 students in all in 6 classes. They had not previously discussed the issue in question.

- During an attempted conversation between a pair of people, designated **X** and **Y**, **X** endeavoured to use as many negative instructions, statements and comments as possible while **Y** kept track of his or her feelings.
- In a trio of people **X** and **Y** endeavoured to have a normal conversation. **Z** monitored events and intervened every time **X** or **Y** used a negative and suggested that the negative be changed into a positive.
- In therapy group or class of students the people took turns to complete sentences starting: "Now I am not aware of …."
- This was given as "home work" as an awareness experiment and, for clarity, is now expressed in first person grammar: "Each time I catch myself using a negative in a sentence I will promptly alter it to the positive form that will get me what I want and will observe the effect on myself and others."

Results

In the first two experiments students reported confusion about what was being communicated in the negative instructions. There was sometimes confusion about the meaning of negative statements. In either case, change to positivity produced accuracy in providing and interpreting information.

In the third experiment they all reported that it was impossible to be not aware because as soon as they became aware that something was out of awareness it entered awareness. There was no confusion reported, only some anger at being given such a "stupid experiment."

In the fourth experiment the students reported –
 i) difficulty in making the change and
 ii) that use of positive statements lead to significant improvement in clarity of communication.

Some encouraged the people around them to also make the change and reported mutual benefit.

Discussion

Observations and results will be discussed at the practical, moment to moment level, before examining the implications of the initial hypothesis for the gestalt approach. It is important to re-emphasize i) that the reported observations and experiments, with two exceptions, did not occur in a therapy context and ii) that concern here is with negative instructions, commands, and imperatives. Negatives in other contexts, such as statements and comments, may not produce a similar degree of confusion, and, indeed, are an essential form of clear communication. "Do you like gin". "No".

The observations with the children and senior citizens showed that they were evidently obeying the unexpressed commands: "Hit him", "Run", "Run in the road" and "Get out of the bath". Each of these was the obscured positive content of what was actually said, and it can be concluded empirically, if not logically, that the positive must be known, even if out of awareness, before it can be negated.

The experiments showed that, on a day-to-day conversational level, everybody concerned reported that, in relation to other people, reformulation of requests to eliminate negatives resulted in improved communication and personal benefit. It became obvious, in class discussion, that abandonment of the use of negative commands was abandonment of a self-defeating verbal behaviour.

The third experiment provided a special situation because the leading statement was equivalent in effect to "Now I am aware … " as if the negation had not been expressed, and was redundant. This again illustrates that the positive content is processed first.

Comment

From a common sense point of view the general conclusion from the examples and the results of the experiments is that to be sure to gain a desired effect it is necessary to avoid the use of negation in commands, and be careful with comments and statements. At the time of each exemplifying incident the perpetrator of the negative had dominating figures operating; "talking to a bank clerk", etc., and was probably interrupted by an unexpressed thoughts, "What's he doing?" The immediate reaction to the interruption was fast, without thought, and was probably related to past behaviours. These behaviours may have been introjects operating or may have been decision based. In either case the new figure, a negative command, failed to have the desired effect.

For future use the introject or established decision involving the negation could be subject to a redecision processes which would also be, in the therapeutic context, a curative move (Goulding & Goulding[8]). That is if the person is willing to discuss or wants to make such a new decision.

The redecision involved is a change of behaviour that can be expressed as: "In the future I will only make positive statements" and "Each time I catch myself using a negation I will change it and say what I want in positive form". The latter manoeuvre can only succeed in a sharp state of awareness and, in the early days of carrying out the experiment, may result in much interruption of communication with other people.

To make the positive instruction a completely different verbal formulation is needed: "Don't throw" becomes "Eat your food". Similar transformations are, in example 2: "Be gentle"; in 3: "walk slowly"; in 4: "Stand still".

The anecdotal information provided here may take adverse comment on several counts. For example, in the third example there was a change of person giving the message. The second teacher may have had a more effective authority over the children. Another, paradoxical issue, which could be discussed elsewhere, has appeared in the last example; memory can only be improved if one remembers to do so.

There is historical support for the avoidance of negative verbal expressions. Thus Coué[10], in advocating the use of affirmations, stressed the need for each to be positive statements though he omitted discussion of reasons. Baudouin[11] did so discuss and emphasized the obvious improvements of interpersonal relationships that follow the change to positivity.

Engagement and action

In being concerned about the gestalt approach my primary interest is to examine the closeness of the relationship of the involved people. Does the use of negative commands enrich, have no effect on, or disrupt communication, contact and mutual action? The answer to these questions, based on the observations and results of experiments reported above, is that dramatic misunderstanding and confusion can occur when negative commands are given. It is now appropriate to assimilate this information in an analytical, gestalt context.

An examination of interruption and avoidance processes can start with the proposition that the person using negatives may be operating from an introject, projecting, deflecting or is demanding confluence. However, the most obvious aspect is the expression of polarity positions as stressed above. Another way of envisaging disruption of figure development is to consider the hypothesis that it is the use of a negative that sets up the confusion producing polarity; "run" / "don't", "get out of the bath" / "don't". This is so in spite of the fact that the processes under discussion occur out of awareness of the person concerned

and were probably learned from parents and teachers.

Each example was presumably accompanied by a fantasy for the user of the negative; "He's hitting him", "He'll get killed if he runs in the road", etc., fantasies prompted by old introjects or learning. These become projections that are potentially valuable as the purpose can be considered to be avoiding hurtful activity or accident. The nub of this issue is that the actual erroneous language form produced the opposite effect to that desired.

Further discussion based on figure/ground theory pays attention to the stages of figure development and particularly to the consummatory contact and action stages where the original need is met. The need was not met when negation occurred and thus the appearance of a negation in a sentence indicates interruption of healthy figure development. Such a figure is only viable if the recipient of the negative command can decode it and this may imply having sufficient time to do so.

Therapy

As a therapist I reckon to be particularly sensitive to my own and other people's interruptions to inter-personal engagement. In group I share my awareness. To one client I said: "You said, very energetically, 'don't confront me'. I now feel confused because I hear and feel most strongly the positive content of your command which is 'confront me'. What do you really want?" The reply was: "I want your attention and yet I am embarrassed when you show me up in front of these other people".

To put it another way, my task is to share my awareness of interruption of figure development and of the origin of the confusion in such a way that the person concerned can be aware of the process and, if he or she decides to do so, make a change so that figure completion occurs. Negatives can be effectively negated only by reformulating into a new positive statement or command using different wording, as was discussed above. This is required for the move towards healthy figure formation although, in the therapy situation, the client had best discover this for him or her self. Concern is then with ways of reconstructing figure development.

There were also two complications. First, it was easy to recognize the negation in: "Do not throw your food around". There was more problem with "don't" because the "n't" at the end of the word tended to become swallowed. Of particularly vicious effect were "shouldn't", "mustn't", "oughtn't", etc.

For me, in the therapy or other situation, I know that I communicate clearly and effectively when I model the beneficial behaviour; I intend to never use negative commands.

At one time many men sang often about the Derby Ram: "And didn't he ramble ..." and all present had no doubt that he did indeed ramble extensively:" ... ""till the butcher cut him down".

Go to the edge
Dance on the edge
Return fortified.

Notes and references

1   These observations on HEALTH originated with Dr T. L. Nichols who was concerned about the medical obsession with the signs and symptoms of disease. His points have been abridged for use at the beginning of the 21st century. These ideas on health were presented and discussed with refreshing alacrity by Diamond[13].

2   Cluff, 1984.

3   For a right handed person the dominant hand is the right hand so the tonic is the left: a musical metaphor to help your recognition.

4   My memory tells me that Dave Dobson taught me this procedure.

5   Procedure adapted from "Progressive Relaxation" by Jacobson[12] See also p. 163 of Green & Green[14], and Perls, Hefferline & Goodman[15], When working with cancer patients I taught them six relaxation methods and the Jacobson procedure was judged by them to be the most effective.

6   I vaguely remember coming across this procedure in a Buddhist text.

7   This is *The Old House* guided fantasy that has appeared in many places and is now updated in two prominent respects – discovery and use of the observatory and using a way back that differs from the outward journey.

8   Goulding & Goulding, 1979, pp. 189 → 193.

9   Polster and Polster, 1973.

10   Coué, 1915.

11   Baudouin, 1920.

12   Jacobson, 1923 and 1976.

13   Diamond, 1979.

14   Green and Green, 1977.

15   Perls, Hefferline and Goodman, 1951.

—=(✩)=—

References

Anderson, R. (1999). *Human Evolution, Low Back pain, and Dual-Level Control*. In Trevathan, Smith and McKenna, (1999).

Ansbacher, H.L. & Ansbacher, R.L., Eds., (1964) *The Individual Psychology of Alfred Adler; A Systematic Presentation in Selection from his Writings*. Harper, New York.

Ash, M.G. (1995). *Gestalt Psychology in German Culture; Holism and the Quest for Objectivity*. C.U.P.

Barber, P. (2006), *Becoming a Practitioner Researcher: A Gestalt Approach to Holistic Inquiry*. Middlesex University Press, London.

Barnard, C. (1949). *Introduction to the Study of Experimental Medicine*. New York.

Baudouin, C. (1920). *Suggestion and Autosuggestion; A psychological and pedagogical study based upon the investigations made by the new Nancy school*. Allen & Unwin, London.

Benedek, T. 1952. *Psychosocial function in Women*. Ronald, New York.

Berne, E. (1972). *What do You Say after You Say Hello? The Psychology of Human Destiny*. Grove Press, reprinted 1962, Corgi, London.

Berne, E. (1973). *Transactional Analysis in Psychotherapy: A Systematic Individual and Social Psychiatry*. Balantyne, New York,. Reprint of 1961 Grove Edition.

Berne, E. (1976). *Beyond Games and Scripts*, Grove Press, New York.

Brostoff, J. and Gamlin, L. (2008). *The complete guide to Food Allergy and Intolerance*. Quality Health Books, Devon.

Brüne, M. (2008). *Textbook of Evolutionary Psychiatry: The Origins of Psychopathology*. Oxford University Press, Oxford.

Buss, D.M. (2008), *Evolutionary Psychiatry: The New Science of the Mind*, 3rd. Edn., Pearson, Boston.

Cannon, W.B. (1963). *The Wisdom of the Body*. Norton, New York.

Carlock, C.J., Glaus, K.O. & Shaw, C.A. (1992). *The Alcoholic: A Gestalt View* in Nevis, (1992).

Carradoc-Davies, G. (1995). *A Return Journey to the Concept of Top Dog / Under Dog travelling with Winnicot and others*. Brit. Gestalt. J., **4**, pp 129 - 137.

Castaneda, C. (1985). *The Fire from Within*, Black Swan, USA.

Clark, A.J. (1998). *Defense Mechanisms in the Counselling Process*. Sage, London.

Clarkson, P. (1989). *Gestalt Counselling in Action*. Sage.

Cluff, R.A. (1984). *Chronic Hyperventilation and its Treatment by Physiotherapy: Discussion Paper*. J. Roy. Soc. Med., 77, pp 855 → 862.

Coolidge, F.L. and Wynn, T. (2009). *The Rise of Homo Sapiens: The Evolution of Modern Thinking*. Wiley-Blackwell, Chichester.

Cordain, L. (2002). *The Paleo Diet: Lose Weight and get Healthy by Eating the Food You were Designed to Eat*. Wiley, New Jersey.

Coué, E. (1915), *De la suggestion et de ses applications*. Barbier, Nancy.

Coué, E. (1926) reprinted 1981, *Self-Mastery through conscious Autosuggestion*, Sun Books, Santa Fe.

Darwin, C., (1859), *On the Origin of Species*, Murray, London.

Dawkins, R. (2009). *The Greatest Show on Earth: The Evidence for Evolution*. Bantam Press, London.

Deikman, A.J. (1982). *The Observing Self: Mysticism and Psychotherapy*, Beacon Press, Boston.

Diamond, J. (1979). *BK. Behavioural Kinesiology: How to Activate Your Thymus and Increase Your Life Energy*. Harper and Row, London.

Dobson, D. (1981). *Magical Mysteries of the Mind*. In press.

Dryden, W. (1990), Ed., *Individual Therapy: A Handbook*. Open University Press, Milton Keynes.

Dunbar, H.F. (1947) *Mind and Body: Psychosomatic Medicine*. Columbia Press, New York.

Dunbar, R., Barrett, L. and Lycett, J. (1994). *Evolutionary Psychology, A Beginners Guide: Human Behaviour, Evolution and the Mind*. One World, Oxford.

Eaton, S.B. and Eaton, S.B. III. (1999). *Breast Cancer in Evolutionary Context* in Trevathan, Smith and McKenna, (1999).

Edwards, R.W.H. (1984). *Holism and Gestalt*, Brit. Holistic Medical. Assoc. Newsletter, No 5, 1984.

Edwards, R.W.H. (1984b). *Towards N-L.P. with a human face. An appreciation of the first diploma course in neuro-linguistic programming to be run in Europe*. Self and Society; The European Journal of Humanistic Psychology", Vol XII, No 2, 1984, pp 89 → 100.

Edwards, R.W.H. (1985). *Psychosomatic Disease and Health: A Gestalt Therapy Approach*. Masters Thesis, Antioch University, Yellow Springs, Ohio.

Edwards, R.W.H. (1991). *Techniques and Strategy*. Brit. Gestalt Journal, **1**, p 52.

Edwards, R.W.H. (1993a). *Unproductive Breathing*. Brit. Gestalt. J. **2**, pp 131 → 132.

184

Edwards, R.W.H. (1993b). *The theory and practice of Assertion Skills; an applied Gestalt Approach.* Brit. Gestalt. J. **2**, pp 44 → 52.

Edwards, R.W.H. (1994). *An acute breathing disorder.* J.R.S.M., **87**, p 644.

Edwards, R.W.H. (1996). *A Gestalt approach to Food Intolerance.* In Feder & Ronall, (1996).

Edwards, R.W.H. (1998). *Becoming an Ex-Smoker.* Brit. Gestalt. J. **7**, pp 45 → 48.

Edwards, R.W.H. *A Role for Protofigures in Therapy.* Brit. Gestalt J. in press.

Edwards, R.W.H. *Evolutionary Gestalt.* In press.

Ehrenfels, C. D. von (1938), in Ellis, (1938, page 4.).

Ellis, W.D. (1938). Ed. *A Source Book of Gestalt Psychology,* Routledge & Kegan Paul, London.

Enright, J.B. (1970). *An Introduction to Gestalt techniques.* In Fagan & Shepherd (1970a), pp 107 → 124.

Erickson, M.H., Rossi, E.L. & Rossi, S.I. (1976). *Hypnotic Realities: The Induction of Clinical Hypnosis and forms of Indirect Suggestion.* Irvington, New York.

Evans, D. and Zarate, O. (2005). *Introducing Evolutionary Psychology.* Icon Books, Cambridge.

Fagan, J. & Shepherd, I., Eds., (1970). *Gestalt Therapy Now; Theory, Techniques, Applications.* Science and Behaviour Books, Palo Alto and Harper Colophon, New York.

Feder, B. & Ronall, R., Eds. (1980). *Beyond the Hot Seat; Gestalt Approaches to Group.* Brunner/Mazel, New York.

Feder, B. & Ronall, R., Eds. (1996). *A Living Legacy of Fritz and Laura Perls: Contemporary Case Studies.* Feder & Ronall, New Jersey.

Freud, Anna, (1937). *The Ego and Mechanisms of Defence,* Revised Edition, Karnac Books, London.

Gendlin, E.T. (1996). *Focusing-Oriented Psychotherapy; A Manual of the Experimental Method.* Guilford Press, New York.

Gilbert, P. (1992). *Counselling for Depression.* Sage, London.

Goldstein, K. (1963). *The Organism,* Beacon Press, Boston.

Goncharov. I. A. (1873) and (1092). *Oblomov.* Everyman Library.

Gottlib, I.H. & Hammen, C.L. (1992). *Psychological aspects of Depression: Towards. A Cognitive - Interpersonal Integration.* Wiley, Chichester.

Goulding, R. & Goulding, M. (1979). *Changing Lives through Redecision Therapy.* Grove Press, New York.

Green, E. & Green, A., 1977, *Beyond Biofeedback,* Delacorte, USA.

Green, R.L. and Ostrander, R.L. (2009). *Neuroanatomy for Students of Behavioural Disorders.* Norton. New York.

Greenfield, S. (1997). *The Human Brain; A guided tour.* Weidenfeld & Nicholson, London.

Grinker, R.R. (1973), *Psychosomatic Concepts,* Aronson, New York.

Grove, D.J. & Panzer, B.J. (1989). *Resolving Traumatic Memories; Metaphors and Symbols in Psychotherapy,* Irvington, New York.

Grove, D.J. (1989). 8 video tapes, 2 audio tapes and a manual. *Healing the Wounded Child Within.* David Grove Seminars, Edwardsville, USA.

Grove, D.J. (1989). 1 video tape, 4 audio tapes and a manual. *Resolving Feelings of Anger, Guilt and Shame.* David Grove Seminars, Edwardsville, USA.

Grove, D.J. undated, 4 video tapes. *Metaphors to Heal by: A study course in epistemological metaphors.* David Grove Seminars, Edwardsville, USA.

Gut, E. (1989). *Productive and Unproductive Depression; Success or failure of a Vital Process.* Tavistock/Routledge, London.

Hall, R.A. (1976) *A Schema of the Gestalt Concept of the Organismic Flow and its Disturbance.* In Smith (1976a), pp 53 → 57.

Hare, D. (1997). Translator of *Ivanov* by Anton Chekhov, Methuen Dramas, London.

Horney, K. (1945). *Our Inner Conflicts: A Constructive Theory of Neurosis.* Norton, New York.

Jacobson, E. (1976). *You must Relax.* 5th edition. McGraw-Hill, New York. (1st edition, 1923)

Janet, P. (1924). *Principles of Psychotherapy.* Trans, H.M. & E.R. Guthrie, Allen & Unwin, London.

Johnstone, E.C. Ed., (1996). *Biological Psychiatry.* British Medical Bulletin, 52, No 3.

Kandel, E.R., Schwartz, J.H. and Jessell, T.M. Eds. (2000). Principles of Neural Science, 4th Edn., McGraw-Hill, New York.

Kepner, J.I. (1987). *Body Process; A Gestalt Approach to Working with the Body in Psychotherapy.* Gestalt Institute of Cleveland Press, New York.

Koffka, K. (1935). *Principles of Gestalt Psychology.* Harcourt, Brace & World, New York.

Koffka, K. (1938). *Introduction.* In Ellis, (1937).

Köhler, W. (1947). *Gestalt Psychology,* Liveright, New York.

Kopp.R.R. (1995). Metaphor Therapy: Using Client-Generated Metaphors in Psychotherapy. Brunner-Routledge, New York.

Kuhn, T.S. (1996), 3rd Edn., The Structure of Scientific Revolutions. University of Chicago Press, Chicago.

LaFrenière, P., (2010). *Adaptive Origins: Evolution and Human Development.* Psychology Press, New York

Laign, R.D. (1960). *The Divided Self.* Penguin, Harmondsworth.

Lankton, S.R. & Lankton, C.H. (1983). *The Answer Within: A Clinical Framework of Ericksonian Hypnotherapy*, Brunner/Mazel, New York.

Latner, J. (1992). *The Theory of Gestalt Therapy* in Nevis, (1992).

Lawley, J. & Tompkins, P. (2000). *Metaphors in Mind: Transformation through Symbolic Modelling*. Developing Company Press and Anglo-American Book Co.

LeDoux, J. (1998). *The Emotional Brain; The Mysterious underpinnings of Emotional Life*. Weidenfeld & Nicolson, London.

Levitsky, A. & Perls, F.S. (1970). *The rules and games of gestalt therapy*. In Fagan & Shepherd, (1970a), pp 140 → 149.

Lewin, K. (1937). *Will and Needs*. In Ellis, (1937).

Lewin, K. (1948). *Field Theory in Social Science*. Harper & Row, New York.

Marcus, E.H. (1979). *Gestalt Therapy and Beyond: An Integrated Mind–Body Approach*. Meta Publications, Cupertino, California

Maslow, A.H. (1967). *A theory of Metamotivation: The Biological Rooting of the Value-life*. J. Humanistic Psychology, Fall, 93 → 127.

McCabe, J. (1948). *A Rationalists Encyclopaedia*, Watts, London.

McGuire, M. and Troisi, A. (1998). *Darwinian Psychiatry*. Oxford University Press, Oxford.

Melnick, J. & Nevis, S.M. (1992) *Diagnosis: The Struggle for a Meaningful Paradigm* in Latner, (1992).

Miller, G.A. (1956). *The Magical Number Seven plus or minus Two: Some limits on our capacity for processing information*. Psych. Rev. **63**, 81 – 97.

Mitchell, J. (1996). *Bill: Finding roots in the hidden Ground of Loss*. in Feder & Ronall, (1996).

Mollon, P. (1996). *Multiple Selves, Multiple Voices; Working with Trauma, Volition and Dissociation*. Wiley, Chichester.

Murray, D. (1995). *Gestalt Psychology and the Cognitive Revolution*. Harvester, New York.

Naranjo, C. (1970). *Present Centredness: Technique, Prescription and Ideal*. In Fagan & Shepherd, (1970a), pp 47 → 69.

Nesse, R.M. and Williams, G.C. (1994). *Why We Get Sick: The New Science of Darwinian Medicine*. Times Books / Random House, New York.

Ockham, 1646. Much quoted and interpreted elsewhere.

Parlett, M. (1996). *Editorial*, Brit. Gestalt J., **6**, p 2.

Parlett, M. & Page, F. (1990). *Gestalt Therapy* in Dryden (1990).

Perls, F., (1944). *Ego, Hunger and Aggression*, South African edition.

Perls, F., (1947). *Ego, Hunger and Aggression; The Beginning of Gestalt Therapy*. Allen & Unwin, London. Reprint of Perls, (1944). See also Perls, (1969).

Perls, F., (1948). *Theory and Technique of Personality Integration*. Am. J. Psychotherapy, 2, pp 565 → 586. Reprinted in Stevens, (1975), pp 44 → 68.

Perls, F., (1949). *In and Out the Garbage Pail*. Real People Press, Lafayette, Calif.

Perls, F., (1955). *Morality, Ego Boundary and Aggression*. In Stevens, (1976), pp 27 → 38.

Perls, F., (1966). *Four lectures*. In Fagan & Shepherd, (1970a), pp 14 → 38.

Perls, F., (1967). *Group vs. Individual Therapy*. Reprinted in Stevens, (1975), pp 9 → 15.

Perls, F., (1969). *Gestalt Therapy Verbatim*. Real People Press, Lafayette, Calif.

Perls, F., (1969). *Ego, Hunger and Aggression; The Beginning of Gestalt Therapy*. Reprint of Perls, (1947). Vintage, New York.

Perls, F.S. (1970). See Fagan & Shepherd, (1970).

Perls, F., (1973). *The Gestalt Approach and Eye Witness to Therapy*. Science and Behaviour Books, Palo Alto. Reprinted by Bantam, 1976.

Perls, F., (1975a). *Resolution*. In Stevens, (1975), pp 70 → 75.

Perls, F.S. (1975b). *Biography* see Shepherd (1975).

Perls, F., Hefferline, R.F. & Goodman, P. (1951). *Gestalt Therapy; Excitement and Growth in the Human Personality*. Julian Press, New York. Reprinted by Bantam, New York, 1977.

Perls, F. (1975c). *Legacy from Fritz; Lectures, Memories and Transcripts*. In Baumgarden, (1975). Pp 89 → 218.

Perls, L. (1953). *Notes on the Psychology of Give and Take*. Complex. Reprinted in Pursglove, (1968), pp 118 → 128.

Perls, L. (1956). *Two Instances of Gestalt Therapy*: Case Reports in Clinical Psychology, (1968), **3**, pp 109 → 146, Reprinted in Pursglove, (1968), pp 42 → 63.

Perls, L. (1961). *One Gestalt Therapists Approach*. Annals of Psychotherapy, Vols 1 & 2. Reprinted in Fagan & Shepherd, (1970a), pp 125 → 129.

Perls, L. (1976). *Comments of the New Directions*. In Smith, (1976a), pp 221 → 226.

Perls, L. (1992). *Living at the Boundary*. Ed. Wysong, J. Gestalt Journal Publication, New York.

Perls, L. (1994a). *Concepts and Misconceptions of Gestalt Therapy*, in Smith (1994), pp 3 → 8.

Perls, L. (1994b). *A Trialogue between Laura Perls, Richard Kitzler and E. Mark Stern*, in Smith (1994), pp 18 → 32.

Philippson, P. (2009). *The Emergent Self; An Existential-Gestalt Approach*. UKCP/Karnac, London.

Pinker, S. (1994). *The Language Instinct; The New Science of Language and Mind*. Penguin, Harmondsworth.

Pinker, S, (1997). *How the Mind Works*, Alan Lane, The Penguin Press, Harmondsworth.

Plotkin, H. (1997). *Evolution in Mind; An Introduction to Evolutionary Psychology*. Penguin, Harmondsworth.

186

Plotkin, H. (2004). *Evolutionary Thought in Psychology: A Brief History*. Blackwell, Oxford.

Polster, E. & Polster, M. (1973). *Gestalt Therapy Integrated: Contours of Theory and Practice*. Vintage and Brunner Mazel (1974), New York.

Popper, K.R. (1959). *The Logic of Scientific Revolutions*.

Potter. S. (1950). *Some Notes on Lifemanship*. Penguin, Harmondsworth.

Reich, W.R. (1960). *Selected Writings: An Introduction to Orgonomy*. Vision, London.

Reich, W.R. (1960). *The Function of the Orgasm: Sex-economic Problems of Biological Energy*. Panther, London. Reprint of the 1942 edition.

Rose, A. (1990). *Gestalt and Long Term Therapy*. Gestalt South-West News Letter.

Rose, H. and Rose, D. Eds., (2000). *Alas Poor Darwin: Arguments Against Evolutionary Psychology*. Cape, London.

Rothschild, B. (2000). *The Body Remembers: The Psychophysiology of Trauma and Trauma Treatment*. Norton, New York.

Rowe, D. (1983). *Depression: The way out of your Prison*. Routledge, London.

Rubin, E. (1914) in Ash (1995), page 179.

Schoen, G. (1994). *Presence of Mind; Literary and Philosophical Roots of a Wise Psychotherapy*, Gestalt Journal Press, New York.

Schulte, H. (1924) in Ellis, W. D. (1938), pp 362 → 369.

Schwartz, J.H. (2000). *Consciousness and the Neurobiology of the Twenty-First Century*. in Kandel, E.R., Schwartz, J.H. and Jessell, T.M. Eds. (2000). p 1317.

Selye, H. (1952). *The Story of the Adaptation Syndrome*. Acta, Montreal.

Selye, H. (1956). *The Stress of Life*. McGraw-Hill, London.

Simkin, J.S. (1976). *Gestalt Therapy Mini-lectures*. Celestial Arts, California.

Sluckin, A. (1993). *Addressing the Whole Field: The Implication of Food and Chemical Intolerance for Psychotherapy*. Brit. Gestalt J., **2**, 10 → 18.

Smith, E.W.L. (1996). Ed. *Gestalt Voices*. Abbex, New Jersey.

Smuts, J.C. (1926). *Holism and Evolution*. MacMillan, New York; Reprinted (1996), Ed. Wysong, Gestalt Journal Press, New York.

Snow, C.P. (1966). The Two Cultures. McMillan, London.

Soloman, I. (1995). *A Primer of Klienian Therapy*. Aronson, New York.

Stern, D.G (1995). *Wittgenstein on Mind and Language*. OUP, Oxford.

Stevens, J. (1967). *Don't Push the River*. Bantam, New York.

Stevens, J. (1976). *Awareness*. Bantam, New York.

Swanson, J. (1980). *The Morality of Conscience and the Morality of the Organism; Valuing from the Gestalt point of View*. Gestalt J., III, No 2, pp 71 → 85.

Tillett, (1984). *Gestalt Therapy in Theory and Practice*. B. J. Psychiatry. **145**, 231 → 235.

Tomasello, M., (2009). Why we Cooperate. A Boston Review Book, Cambridge, Mass.

Trevathan, W.R, Smith, E.O. and McKenna, Eds., (1999). Oxford University Press, Oxford.

Tutu, D. M. (1999). *No Future without Foregiveness*. Rider, London.

Wertheimer, M. (1945). *Productive Thinking*. Univ. of Chicago Press.

Wertheimer, M. (1938). in Ellis (1938)

Wertheimer, M. in Murray, (1995).

Wertheimer, M. in Ash. (1995).

Westen, D. (1998). *Psychology: Mind, Brain and Culture*. 2nd Edn. Wiley, Chichester.

Wheeler, G. (1991). *Gestalt Reconsidered; A New Approach to Contact and Resistance*. Gestalt Institute of Cleveland Press and Gardner Press, New York.

Wingfield, E. (1999), *Performance Anxiety in Classical Singers and Musicians – A Gestalt Perspective*. British Gestalt Journal, **8**, pp 96 → 108.

Winnicott, D.W. (1971). *Playing with Reality*. Routledge, London.

Wolpe, J. (1988). *Panic disorder: A product of classical conditioning*. Behaviour Research and Therapy, **26**, 441 → 450.

Wymore, J. (2006). *Gestalt Therapy and Human Nature: Evolutionary Psychology Applied*. Authorhouse, Illinois.

Yontef, G. (1991). *Techniques in Gestalt Therapy*. British Gestalt Journal, **1**, p 114.

Yontef, G. (1993). *Awareness, Dialogue and Process: Essays on Gestalt Therapy*. Gestalt J. Press, New York.

Yudofsky, S.C. (2005). *Fatal Flaws: Navigating Destructive relationships with people with Disorders of Personality and Character*. American Psychiatric Publishing, Inc., Washington, D.C.

Zeigarnik, B. (1927). *On Finished and Unfinished Tasks*. In Ellis, (1938), p 300 → 314

Zinker, J. (1977). *Creative Process in Gestalt Therapy*. Vintage Books, New York.

—=(☆)=—

Index

—=(☆)=—